# Psychological Science and Christian Faith

# Psychological Science
*and* Christian Faith

## INSIGHTS AND ENRICHMENTS FROM
## CONSTRUCTIVE DIALOGUE

Malcolm A. Jeeves
*and* Thomas E. Ludwig

TEMPLETON PRESS

Templeton Press
300 Conshohocken State Road, Suite 500
West Conshohocken, PA 19428
www.templetonpress.org

Set in Minion Pro by Gopa & Ted2, Inc.

Library of Congress Control Number: 2018935606
ISBN: 978-1-59947-522-6 (cloth: alk. paper)

This paper meets the requirements of ANSI/NISO Z39.48-1992
(Permanence of Paper).

A catalogue record for this book is available
from the Library of Congress.

Printed in the United States of America.

22  21  20  19  18     10 9 8 7 6 5 4 3 2 1

To Ruth and Deb, for their love and support.

To Kathryn Brownson and Kayla Pouliot.
Your assistance has been invaluable.

# Contents

Preface      ix

CHAPTER 1    Resetting the Agenda      3

CHAPTER 2    The Conflict Motif in Historical Perspective      21

CHAPTER 3    From Conflict to Concordism      43

CHAPTER 4    Integration under the Microscope:
Historical Perspective      61

CHAPTER 5    Integration: Contemporary Views      77

CHAPTER 6    Insights from Neuropsychology: An Overview      99

CHAPTER 7    Insights from Neuropsychology about Spirituality      125

CHAPTER 8    Insights about Conversion, Morality,
Wisdom, and Memory      151

CHAPTER 9    Insights from Evolutionary Psychology      173

CHAPTER 10    Insights about Human Needs and Motivation      193

CHAPTER 11    Social Psychology and Faith:
Stories of Conflict, Concordism,
and Authentic Congruence *(by David G. Myers)*      209

CHAPTER 12    The Way Ahead for Psychological Science
and Christian Faith      229

References      245

Name Index      275

Subject Index      283

# Preface

GIVEN PSYCHOLOGY'S POSITION as the "science of behavior and mental processes," in any discussions of the relationship between psychology and Christian faith we should consider what can be learned from the other sciences. Fortunately, historians have documented the shifting relationship between religion and the various scientific disciplines across many centuries. Indeed, if one also takes into account science's forerunner in the form of "natural philosophy," it is possible to trace these changing relations over two millennia. This historical research demonstrates that scientists and theologians have both made errors along the way, and there are lessons to be learned from those errors. For example:

- Scientists reached premature conclusions before all the relevant evidence was examined, or made sweeping claims that went beyond the evidence.
- Theologians made reflexive reactions against new scientific discoveries that appeared to challenge familiar religious concepts.
- Scientists and theologians both constructed "concordisms" that tried to harmonize current scientific theories and currently accepted interpretations of scripture.

The lessons learned by other sciences and from earlier generations of Christians should help us avoid making similar errors as we work to formulate a constructive relationship between psychological science and Christian faith. The purpose of this book is to promote progress toward that constructive relationship. Throughout this book we attempt to show the inadequacy and inaccuracy of the *conflict motif* and the dangers of premature *concordism*. We highlight examples of an alternative approach—often called *complementary perspectives*—that emphasizes the *mutual insights* and *enrichments* that can emerge from a cordial working relationship between psychology and theology.

Why did we choose to attempt this task? One of us (MAJ) has been privileged to be involved personally in some of the major scientific developments in psychology over more than half a century, including

- The onset and development of the so-called cognitive revolution
- The development of neuropsychology
- The rapid advances in evolutionary psychology

These personal connections are described more fully in the paragraphs below—first for Malcolm Jeeves and then for Thomas Ludwig.

**The cognitive revolution.** According to Howard Gardner (Gardner 1985), the cognitive revolution began at a meeting held in Cambridge, England, in 1956, called by Sir Frederic Bartlett and jointly organized with Jerome Bruner of Harvard. Malcolm was privileged to be a participant in, and the secretary of, this international conference. Thus, he has been involved in cognitive psychology from the very beginning and has followed developments in this area ever since, as well as publishing research on thinking and structural learning, which were central themes of the cognitive revolution (e.g., Dienes and Jeeves 1965; Dienes and Jeeves 1970; Jeeves and Greer 1983). As one indication of the impact of these publications, they have been translated into French, German, Hungarian, and Italian.

**Neuropsychology.** Malcolm has been active in research in neuropsychology and cognitive neuroscience for more than fifty years (e.g., Jeeves 1965; Jeeves and Baumgartner 1986; Lassonde and Jeeves 1994; Milner and Rugg 1995; Jeeves et al. 2001). For part of that time he served as editor in chief of *Neuropsychologia*, the major scientific journal reporting research in this area. That alone ensured that he was consistently up-to-date as reports of the latest exciting developments worldwide crossed the editor's desk seeking publication.

**Evolutionary psychology.** As well as initiating and publishing work in comparative psychology and evolutionary psychology, including carrying out some of the first systematic studies of marsupial behavior (e.g., Rajalakshmi and Jeeves 1965; Jeeves and Winefield 1969), Malcolm has been privileged to be a member of a university department with an international reputation in evolutionary psychology. This has provided regular opportunities to participate in seminars by world leaders in evolutionary psychology and to keep up-to-date with new developments in the field.

Thomas Ludwig has been involved personally in research on a variety of topics in cognitive psychology and cognitive neuroscience, including hemispheric specialization for verbal and spatial skills, emotion processing, and the physiology of forgiveness (e.g., Ludwig 1982; MacKay and Ludwig 1986; Ludwig and Jeeves 1996; Witvliet, Ludwig, and Vander Laan 2001). In addition, for four decades he has been at the forefront of developing computer-based experiments and simulations to facilitate the teaching of psychological science to college students (e.g., Ludwig [1986] 2015; Ludwig 2002; Ludwig and Perdue 2005). His work has emphasized the ways in which psychological theories can and must be put to empirical testing if the science is to move forward on a firm foundation.

Thus, we write about what we know firsthand, as actors and participants rather than as spectators. After examining the conflict model, attempts at concordism, and attempts to integrate psychology and religion, we provide examples of complementary perspectives from the areas of psychological science in which the two of us have been directly involved for decades: neuropsychology, cognitive psychology, and evolutionary psychology.

The book's final chapters provide additional examples of complementarity (as well as examples of conflicts and concordisms) from other areas of psychology with which we have less direct contact, including social psychology and personality psychology. In order to be as comprehensive as possible in our discussions of the relations of psychological science and Christian faith, we have invited one of today's leading social psychologists, David Myers—who is also the author of the best-selling general psychology textbook—to contribute a chapter on social psychology in which he calls for "authentic dialogue—a dialogue between the emerging insights of science and biblical scholarship."

Each of the research specialties mentioned here has changed as it has developed over the decades, highlighting the danger of assuming that what science is revealing to us today is set in stone. This progressive development of science is directly relevant to one of the main themes of this book, namely, the futility of constructing a concordism or harmonization between today's science and today's most widely accepted interpretations of scripture. Thus, on the basis of centuries of historical evidence in other sciences and a century of developments in psychology, we shall argue that we need to move away from both conflicts and concordisms. We shall also argue that, although the attempt to *integrate* psychology and Christian faith has seemed a worthy goal, the time has come to minimize the use of that term. Instead, we should focus on the *insights* and *enrichments* biblical theology can provide for psychological research and theory, as well as on the *insights* and *enrichments* that psychological science can offer to theology, and to our understanding of scripture, of ourselves, and of the world in which we live.

We should note that the general approach taken in this book resonates strongly with and develops further the approaches advocated recently by theologian and historian Alister McGrath and his wife, psychologist and Anglican priest Joanna Collicutt McGrath. The potential for mutual enrichments (described by McGrath) and the relevance of academic psychology to many science and faith issues (emphasized by Collicut) are central themes of this book, and we are indebted to them both.

# Psychological Science and Christian Faith

# CHAPTER 1

# Resetting the Agenda

## What Sort of Psychology Are We Seeking to Relate to Christian Faith?

WHY *Psychological Science and Christian Faith*? Why not simply *Psychology and Christian Faith*? There is a substantial literature on the relationship between psychology and Christian faith that expresses a wide range of views about the nature and scope of contemporary psychology. For example, two influential edited volumes on this topic are *Psychology and the Christian Faith: An Introductory Reader* (S. Jones 1986) and *Psychology and Christianity: Five Views* (E. Johnson 2010). Although these two books were published almost twenty-five years apart, both have guided discussions of the relationship between psychology and Christian faith in North America. What is striking about the content of these volumes is that they barely mention several areas of contemporary psychological research that have been among the most intensively researched, widely reported, and generously funded over the past six decades, including

- Cognitive psychology
- Neuropsychology and cognitive neuroscience
- Evolutionary psychology

Why are these omissions regrettable? Howard Gardner has documented the crucial role of the cognitive revolution in modern psychology—as the foundation for a large segment of contemporary research, as well as the entire field of artificial intelligence (Gardner 1985). The importance of cognitive neuroscience was emphasized by the US government in its declaration that the last decade of the twentieth century should be labeled the "Decade of the Brain." Likewise, the first decade of the twenty-first century has been labeled the "Decade of the Mind." The impact of evolutionary psychology is evident from the explosive growth in research funding and publications in this field. The minimal coverage of these research areas in the two widely read volumes mentioned above demonstrates a serious mismatch between what is happening at the cutting edge of psychology and what is included in the typical science-faith discussions. We believe that any current and future discussions of psychology and Christian faith should take note of psychology *as it is today* and *as it is taught to today's college and university students.*

One way of finding out what contemporary psychology looks like is to examine the contents of a current standard textbook. For the past three decades, the most widely read textbook of psychology in North America is the one written by David Myers, now in its twelfth edition (Myers and DeWall 2018). In this book, the prologue and chapter 1 deal with the story of psychology and research methodology. The headings of the remaining chapters are as follows:

Chapter 2 The Biology of Mind
Chapter 3 Consciousness and the Two-Track Mind
Chapter 4 Nature, Nurture, and Human Diversity
Chapter 5 Developing through the Lifespan
Chapter 6 Sensation and Perception
Chapter 7 Learning
Chapter 8 Memory
Chapter 9 Thinking and Language
Chapter 10 Intelligence
Chapter 11 What Drives Us: Hunger, Sex, Friendship,
    and Achievement

Chapter 12 Emotions, Stress, and Health
Chapter 13 Social Psychology
Chapter 14 Personality
Chapter 15 Psychological Disorders
Chapter 16 Therapy

Given this listing of major topics in contemporary psychology, it is interesting to examine the academic specializations of the contributors to the two edited volumes mentioned above. The Stanton Jones (1986) volume contains chapters from eleven contributors—one distinguished brain scientist (Donald MacKay), three social psychologists, two psychotherapists, and four philosophers or theologians. Even on the most generous assessment, that volume cannot claim to be covering the current issues at the interface of contemporary psychology and Christianity. To put it bluntly, the book ignores more than three-quarters of contemporary psychology, in the sense that those topics are not represented by contributions from researchers or practitioners actively involved in these areas.

The more recent book edited by Eric Johnson (2010) fares little better. It has seven chapters; the first is a brief history of Christians in psychology and the final chapter discusses the five views offered in the intervening chapters. The five substantive chapters have a total of seven contributors—a social psychologist, a physiological psychologist (now studying personality and religious coping), a clinical psychologist, two counseling psychologists, a philosopher, and a theologian. Once again, if we compare the content of a general psychology textbook with the academic specializations and research areas of the contributors, it appears that *about 80 percent of contemporary psychology is not represented.* This failure to engage with psychology as it is today is highlighted in David Myers's response to the chapter titled "A Christian Psychology View" written by Robert C. Roberts and P. J. Watson. In that response, Myers lists some important findings of psychological science that are not addressed in scripture or the writings of Christian theologians, such as:

- The functions of our two brain hemispheres
- The quantified heritability of a multitude of traits
- The remarkable cognitive abilities of newborns
- The extent to which peer influences trump parental nurture in shaping children's language, smoking habits, and lifestyle choices
- The effects of experience at different ages on the brain's neural networks
- Changes in mental abilities with aging
- How eyewitnesses construct, and reconstruct, memories
- The powers and perils of intuition
- The components of intelligence
- The effects of stress on the body's immune system
- The ways our self-concept guides our information processing
- The effects of aerobic exercise on mild depression and anxiety
- The things that do and don't predict human happiness

Myers then adds, "If establishment psychology is indeed where significant discoveries and new understandings are emerging, do we really want to run off into the corner to create our own Christian psychology? By doing so, do we not risk irrelevance?" (Myers 2010b, 180–81). To restate our rationale for the title of this book, we believe that discussions of psychology and Christian faith should engage psychology *as it is today* and *as it is taught to today's college and university students*—considering the full range of topics within psychological science.

## CHANGING VIEWS OF THE RELATIONSHIP BETWEEN PSYCHOLOGY AND CHRISTIAN FAITH

One common response to the question, "What is the relation between psychology and Christian faith?" is to point to ongoing conflicts between the two. This response is understandable, given the legacies of very influential figures in twentieth-century psychology—such as Sigmund Freud and B. F. Skinner—who considered religion to be incompatible with science. In addition, the antireligious perspective of well-known contemporary scientists such as distinguished evolutionary biologist Richard Dawkins may suggest that there is a necessary and fundamental conflict between psychological science and Christian belief.

The *conflict* motif is not the only one being used today to describe relations between psychology and faith. Another motif might be called the *concordism* motif. This takes a variety of forms, being described as a "harmonization" of psychology and religion, or even as an "incorporation" of psychology into Christian theology. The basic goal of concordists is to show that glimpses of human nature found in scripture "fit together" neatly with descriptions of human nature presented by some psychologists. There is nothing new in this. As we shall document in later chapters, more than a century ago some leading Christian thinkers were championing a brain-mind theory called *phrenology* because of the way it seemed to "fit together" with some Christian beliefs.

But there is another part of the psychology-religion story that is often overlooked. If we look back across the past decades, we find that some Christian psychologists managed to avoid the strident notes of conflict while at the same time keeping well away from the comforting siren voices of concordism. Their approach was characterized by a focus on the insights and enrichments to their faith from their science, and from their science to their faith. For example, David Myers wrote:

> The natural and biblical data are both viewed through human spectacles and are therefore subject to bias and distortion. We should consequently be open to *insights* that come through either nature or Scripture (recognizing God to be the common source of both), while remembering no one's interpretation of nature or Scripture is final truth. This being the case, we may view with some skepticism any attempt to subject theology to science—or science to theology. (Myers 1986, 218, emphasis added)

[7]

In a later section, he makes this focus even clearer, tying the *insights-enrichments* model to a broader approach called *complementarity* or *complementary perspectives*:

> In this essay we will describe how some of the modern *insights* of social psychology (which are drawn from two research areas) illuminate the ancient biblical revelation of our human nature. Social psychologists have their own special *insights* into human nature. Other *insights* into human nature have been independently discerned by biblical scholars. How well do these two realms *complement* one another? My overall answer is that while there are points of tension which deserve our attention, the ancient biblical view of human nature comes home with renewed force as we review the relevant findings of social psychology. (Myers 1986, 218, emphasis added)

As the author of a general psychology textbook, Myers was well aware that textbooks must cover the whole range of psychology, not just a portion representing a few favorite areas of specialization. In the same way, although it might be possible to take one small area of psychology and find ways of harmonizing it with some aspects of Christian faith, that approach severely limits the impact and applicability to today's psychological science in all its scope. For example, someone specializing in personality theory or the study of individual differences may well find that the language they use to describe their research findings is very similar to the ordinary descriptive language used in scripture passages about people's behavior or about human nature. Because of the superficial similarities in language, there would be a strong and understandable temptation to "harmonize" or "integrate" the concepts used in both domains. However, a researcher in neuropsychology, evolutionary psychology, or neuropharmacology would likely use technical language with very clear, precise, and specialized meanings that are only understood in the context of the scientific research being conducted. In that case, there would be little temptation to try to make findings from this kind of research "harmonize" with descriptions about cognition and behavior in everyday language, such as the kind used in scripture. Instead, it would be more natural to adopt an approach based on complementarity or complementary perspectives. We describe that approach in detail in a later chapter, but here we merely remind the reader that the insights–enrichments motif is closely connected to complementarity.

To return to the insights-enrichments motif, the possibility of gaining theological insights from psychological research has been helpfully clarified by Everett Worthington. In a pair of chapters titled "Psychological Science Strengthens Theological Claims" and "Psychological Science Adds New Ideas to Theology," Worthington (2010) identifies several areas in which psychological science affords new insights into theology and enriches our understanding of some traditional theological doctrines—such as the reality of human evil, the dangers of pride and anger, and the benefits of forgiveness. Another person who has supported the insights-enrichments model is Fraser Watts, who points out that enrichment is not a one-way process. Summarizing Watts's approach, Peter Harrison writes:

> Fraser Watts suggests that in the case of psychology and theology a different model is possible. Theology, he contends, can offer special *insights* into the nature of the human person and can thus both critique and enrich psychology. It does so on the first count by contesting overly reductionist explanations of human persons, and on the second, by making contributions that arise out of its special familiarity with such features of human experience as guilt and forgiveness. Watts also demonstrates ways in which psychology can make positive contributions to theology. Here the discussion extends to theological anthropology, biblical hermeneutics, religious experience and

glossolalia. The general model he offers, then, is one in which theology and psychology can be *mutually enriching.* (Harrison 2010, 10, emphasis added)

There is a wealth of literature on the relations between psychology and religion, and it comes as somewhat of a surprise to realize that this literature provides little support for the conflict motif and rarely shows attempts at concordism. In the next section, we summarize some of the important developments. For additional details, see Jeeves (2013).

## LEARNING FROM THE PAST: THE PSYCHOLOGY OF RELIGION

### *Converging Influences*

In the continuing story of the relationship between psychology and faith, the academic study of the psychology of religion stands out because it illustrates both the possibility of conflict between psychology and faith and the possibility of friendly cooperation between psychology and religion as shared partners in a mutual endeavor. Broadly speaking, psychologists who study religion have concentrated their research on what might be called the "roots" of religion (how religious beliefs and behaviors emerge) and the "fruits" of religion (what impact religion has on cognition and behavior).

What are the "roots" of the *psychology of religion* itself? Leslie Hearnshaw identified four influences that converged at the end of the nineteenth century to provide the basis for this field of psychology (Hearnshaw 1964). These were Sir Francis Galton's studies of the manifestations of religion (e.g., prayer); studies by anthropologists, such as Sir James Fraser, of comparative religion and the origins of religions; the writings of theologians, such as W. R. Inge, on mysticism and religious experience; and finally, the beginnings of the systematic study of religious behaviors, best illustrated by Edwin Starbuck's book *The Psychology of Religion* (Starbuck 1899).

## Positive Beginnings

The lines of research mentioned by Hearnshaw laid the groundwork for the psychology of religion, but the field was popularized by the publication of William James's classic work, *The Varieties of Religious Experience* (James 1902). As the title suggests, this book draws attention to the wide range of human experiences of religion and of the life of faith. It reminds us that each person is unique. Any attempt to fit everyone's experience and practice of faith into one mold fails to recognize the wide variety of religious behavior, and also fails to recognize the variety of spiritual experiences of the people described in scripture.

One of James's lasting contributions was his classification of religion into "healthy minded" and "morbid minded" religion. These, he thought, were related to factors of temperament and personality. Paradoxically, the presence of religious faith could energize some individuals to perform acts of care and compassion, while in others the impact of religion could be more negative, producing guilt, depression, and hatred.

William James called attention to individual differences in religion and spirituality. In contrast, James Pratt focused on common themes across religious people and across religious traditions. Pratt's book *The Religious Consciousness* identified stages of religious development from the primitive to the intellectual and the emotional, arguing that these stages were present in all religions (Pratt 1920).

Cambridge psychologist Robert Thouless, a practicing Christian, broadly followed the example set by William James. As set out in *An Introduction to the Psychology of Religion* (Thouless 1923), Thouless's views toward religion were entirely positive. The book examines the factors involved in religious belief, including both unconscious and conscious processes, and devotes special attention to prayer, conversion, and mystical experiences. When the book was reissued in 1981, Thouless indicated that his interest was in religious consciousness rather than religious behavior, and he reiterated his confidence in the reality of the ultimate mysteries lying behind religious faith. Although Thouless was fully aware of the critiques of religion presented by Freud and others, he saw no necessary conflict between psychology and religion, nor did he attempt to construct concordisms to bolster his own religious faith. Instead, his writings are full of deep insights from psychology into religious faith and experience. To read Thouless today is to be enriched in one's own life journey.

### Hostile Perspectives

In contrast to the views of James and Thouless, James Leuba's *A Psychological Study of Religion* (Leuba 1912) exemplified a much more critical perspective on religion. He approached religion from a naturalistic standpoint and contended that the religious life can be explained (or "explained away") in terms of the fundamental principles of psychology. According to Leuba, humans use religion to search for ways of satisfying needs and desires for a better life. Unlike William James, who strongly affirmed the reality of a "higher power" in life, Leuba declared that there is no objective transcendent agent or source connected with a person's religious experience. According to his view, belief in a personal God will, over time, diminish and eventually disappear. Leuba's work provides an early example of the conflict motif to describe the relation between psychology and religion, and foreshadows later attempts by Freud and Skinner to marginalize religion as a relic from humanity's primitive past.

The influential views of Sigmund Freud on the topic of religion may be divided into those on "primitive religion" and those on "developed religion." Throughout both sets of his writings, there is a pervasive theme of covert and, at times, overt criticism of religion, thus supporting the conflict motif. For example, in *Totem and Taboo* (Freud 1919), he attributed the origins of religion to the psychological connection between the "Oedipus complex"—the unconscious hostility that young men supposedly hold toward their fathers—and the "totem" or symbol of group identity as it existed in small primitive groups. Freud suggested that, at one point in the history of a tribal group (a Darwinian "primal horde"), the young men had banded together to kill (and eat) their father in order to possess his wives. According to Freud, this fearful deed was the source of human guilt (the "original sin"). "The totem feast, which is perhaps mankind's first celebration, would be the repetition and commemoration of this memorable, criminal act with which so many things began, social organization, moral restrictions and religion" (Freud 1919, 236). Freud believed that his theory of psychoanalysis provided some clues to the way in which primitive religions developed, especially those that showed some forms of patriarchal totem.

When Freud wrote *Totem and Taboo*, anthropological knowledge was limited. Today, it is clear that many of the so-called scientific facts upon which Freud based his theories were completely incorrect or seriously misleading. In *Moses and Monotheism* (Freud 1939b), he ventured yet further into his speculative theories about the origins of religion. But as anthropological research intensified, Freud's views fell more and more into disrepute. For example, Freud's descriptions of primitive religion were criticized by Bronislaw Malinowski in *Sex and Repression in Primitive Society* (Malinowski 1927) and *The Foundations of Faith and Morals* (Malinowski 1936). In the context of this chapter, it is worth noting that the conflicts that Freud believed he had exposed between religion and science turned out to be *false conflicts*. Freud's views were based on a particular set of anthropological theories that were subsequently revised or abandoned. Just as *false concordisms* become embarrassing to theologians when the scientific findings move beyond them, so also do false conflicts that are built on shaky science.

Moving from primitive religion to developed religion, we find Freud's views summarized in *The Future of an Illusion* (Freud 1934) and in *Civilization and Its Discontents* (Freud 1939a). These books contain Freud's most severe critiques of religion, adding fuel to the conflict motif. For Freud, an "illusion" is any belief system that is based on human wishes and hopes rather than on objective reality. He was careful to point out that such a basis does not necessarily imply that the system is false, but as far as Christianity was concerned, he clearly believed that it was indeed false. He did not deny that religion serves a useful purpose in providing a sense of security in a hostile environment and an important repository and guardian for society's ethical standards. In his view, however, ancient superstitious religions could no longer serve the needs of modern society, so they must be replaced by an alternative, rational foundation for living a civilized life.

In the context of his theory of psychoanalysis, Freud concluded that religion was a temporary "universal obsessional neurosis" that offered a means of escape from the realities of life and a method for coping with the restraints imposed by an organized society. Freud considered religious deities to be "substitute ideal fathers." Thus, religion became the projection of the child's relationships with its earthly father, and "gods" in all their different guises were nothing more than magnified father figures. Freud argued that modern humans must "grow out of" religion in order for society to progress (Freud 1934, 43).

Distinguished behavioral psychologist B. F. Skinner believed that his rigorous scientific approach set him completely apart from Freud and others who theorized about unconscious processes and mental states that could not be observed or measured. Ironically, Skinner's hostile attitude toward religion put him very close to Freud's position on the relationship between science and faith, and very high on the list of individuals responsible for the persistence of the conflict model. Skinner, encouraged by earlier successes achieved using a wide variety of techniques at modifying behavior, speculated about how similar processes could be harnessed to shape the future of society. In his book *Beyond Freedom and Dignity* (Skinner 1971), he proposed that his principles of learning, based on the application of rewards and punishments, offered a possible explanation as to how the practices of religion might function psychologically. Thus, he contended, "a religious agency is a special form of government under which good and bad become pious and sinful. Contingencies involving positive and negative reinforcement, often of the most extreme sort, are codified—for example as commandments— and maintained by specialists, usually with the support of ceremonies, rituals and stories" (Skinner 1971, 116). According to Skinner, the good things personified in a god provide reinforcement for prosocial behaviors, whereas the threat of hell is an aversive stimulus used to shape and control "sinful" behavior. It is clear that a reductionist presupposition underlies Skinner's approach, leaving no room for spiritual influences. Skinner's writings about religion promoted a sense of inevitable conflict between psychology and religion.

## More Balanced Perspectives

Carl Jung worked closely with Freud for a while and developed his personality theory from within the psychoanalytic tradition. It is not surprising that Jung initially agreed with Freud on the significance of the father figure in religion. However, his views eventually changed about both the origins of religion and the impact of religious faith. Jung proposed that there is a "religious instinct" in all humans that leads them to desire a relationship with a higher force or transcendent being. Jung believed that all religions have their psychological roots in what he called the "collective unconscious" of the human race (Jung 1968). Summing up the difference between the Freudian and the Jungian views of religion, G. S. Spinks very aptly wrote that "for Freud religion was an obsessional neurosis, and at no time did he modify that judgment. For Jung, it was the absence of religion that was the chief cause of adult psychological disorders. These two sentences indicate how great is the difference between their respective standpoints on religion" (Spinks 1963, 102). Jung saw no necessary conflict between science and religion. In fact, he believed that science could never replace religion, because religious language and symbolism provide access to deep aspects of consciousness ("archetypes") that are not open to scientific investigation (Jung 1968, 63).

Although less well-known than either Freud or Jung, the distinguished social psychologist and personality theorist Gordon Allport also made a significant contribution to the psychology of religion. His book *The Individual and His Religion* (Allport 1950) has become a classic. In it, he traced the way in which religion develops from childhood and adolescence into maturity, thus underlining the way in which belief in

God functions differently for people at different times in their lives. His account is one of the first to incorporate the results of empirical studies of religious beliefs and behavior. We may regard Allport's book as full of deep insights from a psychological perspective into the role and the function of religion in the lives of individuals and groups. There is no conflict motif here and no attempts at producing concordisms; it is a book full of deep insights into the relationship between psychology and religion.

A psychiatrist whose views received wide publicity in the mid–twentieth century was William Sargant, the author of *Battle for the Mind* (Sargant 1957) and *The Mind Possessed* (Sargant 1974). While writers such as Michael Argyle evaluated different psychological theories of religion against empirical data, William Sargant concentrated on the psychophysiology of religious behavior. Most noteworthy were his attempts to link psychological studies of brainwashing with accounts of religious conversion. He argued that hypnotic suggestion and brainwashing were at work in some large evangelistic campaigns, and identified some of the effective ingredients in such campaigns:

- ► An evangelist with wide publicity and a prestige buildup,
- ► Speaking with great fervor, conviction, and authority
- ► To a crowded meeting
- ► Which began with repetitive singing of emotional hymns and choruses,
- ► Perhaps accompanied by bright lights, large choirs, and stirring music, often with a rhythmic beat.

In such circumstances, argued Sargant, physical and psychological stresses are skillfully applied to produce dramatic changes in both behavior and beliefs. Although Sargant's views are somewhat controversial, we should recognize that he was highlighting important issues that must be considered in Christian outreach and evangelism to ensure that the physical and social environment does not manipulate emotions in a way that makes meaningful judgment and decision-making difficult.

Conversion experiences under such circumstances, sadly, seldom last very long.

In the mid-1980s, Laurence Brown edited an important volume under the title *Advances in the Psychology of Religion* (L. Brown 1985). This book reported on a conference aimed at identifying new directions in the psychology of religion. Brown noted that most of the current research relied heavily upon correlational analysis, while the participants agreed that more experimental methods should be used in future psychological studies of religion. Around that time, developments in cognitive psychology were beginning to produce new ways of studying how people think about God (Brown and Forgas 1980), as well as attempts to identify actual differences between the behaviors of religious and nonreligious people, especially in terms of ethical judgments and moral behavior. Brown's book, and the underlying research it contained, demonstrated that real progress could be made by putting aside the conflict model and focusing on what insights psychological science could shed on religious behavior.

In the context of this book, here is the message we have drawn from our historical review of the psychology of religion: While some researchers and theorists have promoted the conflict motif in the relation between psychology and faith, many others have demonstrated that a cordial, productive relationship is possible. Thus, although some prominent psychologists (most notably, Freud and Skinner) have tried to use psychological principles to "explain away" religious beliefs and behaviors, such a view is not universally accepted by psychologists. For example, Michael Argyle states quite categorically that "it does not follow that because a belief has psychological roots it is therefore false," and that "there needs to be no relation between a psychological basis for belief and the truth of that belief" (Argyle, 1, 4). In short, well-designed psychological studies of religion should not presume the truth or falsehood of religious claims, but rather seek *insights* into the spiritual beliefs and the religious behaviors that are an important part of everyday life for two-thirds of the world's population (Gallup Organization 2017).

## The Need for a Two-Way Dialogue

Cambridge psychologist Fraser Watts has written most helpfully about the dialogue between psychology and theology. He points out that the dialogue between theology and science has been, as he puts it, "notoriously one-sided," in that the history of the science and faith debates has shown theology much more interested in science than science in theology (Watts 2012, 45). To correct this imbalance, Watts suggests that theologians can make a more intentional effort to engage in dialogue with research psychologists as well as counselors. What contributions could theologians make? Watts argues that theologians can bring biblical insights about the human condition to enrich psychological theories and help formulate new research questions. In this regard, Watts is echoing a pervasive theme of this book, namely, the possibility for *mutual enrichment* in the relations between psychology and faith. Watts explains, "The general point is that theology and psychology offer *complementary perspectives* and elucidate different aspects of phenomena such as forgiveness, each of them contributing to the other so that there is value in bringing them into creative dialogue" (Watts 2010, 194, emphasis added). We believe that this dialogue is itself an enriching process. Alister McGrath, in his 2014 Inaugural Lecture when taking up the Andreas Idreos Professorship of Science and Religion at Oxford University, affirmed this point:

> We should challenge the dominant narrative of our time—the outdated "conflict narrative," sustained more by uncritical repetition than by historical evidence—and replace it with a narrative of enrichment. (McGrath 2015, 12)

## Resisting Reductionism

In addition, Watts (2012) claims that theologians can support the attempts of philosophers to weed out pervasive reductionist tendencies within some branches of psychology. Watts identifies four main forms of reductionism in psychology. Through the influence of neuroscience, there is a tendency to see humans as "nothing but a bundle of neurons." From an evolutionary psychology perspective, there is a tendency to argue that humans are "nothing but survival machines for our genes." The artificial intelligence subfield within cognitive psychology views the human mind as "nothing but a sophisticated computer program." From a social psychology perspective, there is a tendency to view human characteristics and aspirations as "nothing but social constructs." We shall deal with some of these topics in greater detail elsewhere in the book, but here we note our agreement with Watts's general point that theological perspectives can help counter the dangers of unthinking reductionism in psychology.

## Looking Ahead at the Plan for This Book

What is the proper or most helpful way to view the relationship between psychological science and Christian faith? Throughout this book we attempt to show the inadequacy of the conflict motif and the dangers of premature concordism. We will highlight examples of an alternative approach—often called *complementary perspectives*—that emphasizes the mutual insights and enrichments that can emerge from a cordial working relationship between psychology and theology.

We follow this introductory chapter with four chapters providing detailed examinations of the three approaches that have dominated the relationship: the conflict model, attempts at concordism, and attempts to integrate the two domains.

The next four chapters provide examples of complementary perspectives from the areas of psychological science in which the two of us have been direct participants: neuropsychology, cognitive psychology, and evolutionary psychology.

The final three chapters provide additional complementarity examples (as well as examples of conflicts and concordisms) from other areas of psychology with which we have less direct contact, and then summarize what we have learned from this project—lessons to keep in mind as we try to connect what is revealed in scripture about how "wonderfully made" we indeed are (Psalm 139:14) to psychological principles that God has allowed us to discover by the dedicated use of the minds we have been given.

# CHAPTER 2

# The Conflict Motif in Historical Perspective

## PERSISTENT CONFLICT BETWEEN SCIENCE AND RELIGION: REAL OR MYTH?

THIS CHAPTER EXPLORES the idea of a conflict between science and religion, in which religion—specifically Christianity—is in a millennium-long retreat, losing ground to the forces of scientific progress. Where did the conflict motif originate? What has sustained it across the years? What impact has it had? We begin with a famous quote from T. H. Huxley that depicts a fierce battle between science and religion:

> Extinguished theologians lie about the cradle of every science as the strangled snakes beside that of Hercules; and history records that whenever science and orthodoxy have been fairly opposed, the latter has been forced to retire from the lists, bleeding and crushed if not annihilated; scorched, if not slain. (Huxley 1860, 556)

However, this battlefield envisioned by Huxley never actually existed. Peter Harrison explains:

> Much recent writing by historians of science has addressed itself, in various ways, to the popular assumption that throughout history science and religion have been engaged in a

perennial battle. It is now generally accepted by historians that this erroneous view, known as "the conflict myth," was largely the invention of two nineteenth-century controversialists, John Draper and Andrew Dickson White. The basic position is clear enough from the titles of their best-known works, respectively, *History of the Conflict between Religion and Science* (1875) and *A History of the Warfare of Science with Theology in Christendom* (1896). Invented or not, the conflict model would not have endured had it not enjoyed at least a superficial plausibility and if it did not play an important role in the self-understandings of those who perpetuate it. In fact, this model draws support from a number of sources: our present experience of religiously motivated anti-evolutionary sentiments and scientifically motivated atheism; well-known historical cases such as the Galileo affair that seem to exemplify conflict; and the assumption that science and religion are forms of knowledge based upon mutually exclusive foundations—reason and experience in the case of science, and faith and authority in the case of religion. (Harrison 2010, 4; see also Draper 1875; White 1896)

Ronald Numbers (Numbers 2009) traces the beginnings of the conflict motif to an 1845 article in a US newspaper that stated, "Every new conquest achieved by science involved the loss of a domain to religion." But this idea was already in the intellectual milieu much earlier. For example, when Pablo de Felipe examined the early roots of the "flat earth error" (the false generalization that Christians before Columbus held that the earth was flat), he found that it was used by several authors in the eighteenth century to build early versions of the conflict model. We believe that the conflict model is an oversimplification of history. We recently studied the history of science and faith relations and showed that a much more complex and richer story better fits the facts (de Felipe and Jeeves 2017). We reviewed twelve historical episodes of these relationships across two thousand years, which have been used to give the impression of a historical directionality—that is, Christians retreating under the steady forward march of science. However, we discovered that many of these stories take unexpected turns, revealing that the conflict-retreat model is historically inaccurate.

Historians tracing out various science-versus-faith scenarios have typically described four key illustrative episodes: (1) in the ancient/patristic times, the debate over the shape of the earth; (2) in the medieval times, the denial of the existence of antipodeans; (3) in the modern era, the debate about the movement of the earth; and, finally, (4) in contemporary times, the rejection of evolution. In all these cases, we are told that Christian theologians were forced to abandon their formerly held positions and retreat in defeat, recognizing the truth and authority of science over the disputed ground. This uneasy truce presumably lasted only until a new conflict broke out at a new science/faith border.

In the hands of a good narrator, the succession of historical controversies about scientific discoveries is always depicted with two contesting sides, and always with the same side (Christianity) shown defending nonsense views that were later destroyed by scientific evidence. This story is an irresistible and emotionally satisfying narrative about the victorious march of science pushing a defeated religious enemy aside—eventually to fade away and disappear. However, a strong case can be made that more careful research of these often-repeated historical episodes shows a much more complex picture, one that resists these simplistic and neat battleground realignments (de Felipe and Jeeves 2017).

A brief look at each of these four "classical" science-versus-faith clashes reveals lessons we should keep in mind as we seek to understand how best to view relations today between psychological science and Christian faith (see de Felipe and Jeeves 2017 for further details).

## THE ANCIENT/PATRISTIC AGE: CHRISTIAN FLAT-EARTHERS

Those who promote the conflict motif often claim that ancient and medieval Christians held the cosmological view of the sixth-century monk

Cosmas Indicopleustes (also known as Constantine of Antiochia), who described a flat earth located in a chest-shaped universe (Cosmas 1897). However, we can find criticisms of this view from Christians in that same era: the Alexandrian philosopher/scientist/theologian Philoponus (sixth century), the Armenian scientist/mathematician Anania Shirakatsi (also known as Anania of Shirak, seventh century), and Photius, the patriarch of Constantinople (ninth century). To support his position, Cosmas used quotations from several earlier Christian writers, but those individuals were mainly connected with the particular theology of the School of Antioch, which by Cosmas's time had already turned into the stronghold of Nestorianism, well outside the mainstream of Christian thought (Wolska-Conus 1989). Moreover, Cosmas had few disciples. Even though his books survived in the Eastern Mediterranean into medieval times, they went largely unnoticed in the West until they were translated into Latin and printed in the eighteenth century. Thus, it is a distortion of the evidence to claim that Cosmas was the source of commonly held flat-earth beliefs.

The only ancient Christian flat-earth author who was well-known in the West was Lactantius, who in the fourth century mocked the sphericity of the earth, although, interestingly, not on theological grounds (Numbers 2009). Later, in the West, Augustine (fourth–fifth centuries) and Isidore (sixth–seventh centuries) were not completely clear about the shape of the earth, but they never denied the sphericity of the earth explicitly as did Lactantius or Cosmas. Starting with Bede (seventh–eighth centuries), a consistent exposition and defense of the sphericity of the earth emerged in Western Europe and made its way into university teaching (de Felipe 2012).

As we look across the ancient and medieval landscape of the West, the debate about the shape of the earth was not a conflict between Christians and the forces of science, but rather a *dispute within Christian thought*, with the flat-earth viewpoint most strongly linked with heretical Nestorianism rather than with mainstream theologians. Eventually, advances in science settled the issue and established the sphericity of the earth. The lesson learned: no service had been done to theology by claiming to find in scripture an answer to a scientific question—and today that includes questions asked by practitioners of scientific psychology.

## THE MEDIEVAL AGE:
### AUGUSTINE AGAINST THE ANTIPODES

Much more complicated problems were posed by claims about the possible existence of the *antipodeans* (humans who lived on the opposite side of the earth). While the flat-earth view ruled out the existence of antipodeans, a spherical earth opened the possibility of dry land on the other side of the earth that could contain a human population. The most common model of a symmetrical continent on the other side of the earth can be traced to Crates of Mallus (second century BCE), who proposed the existence of four land masses, one in each of the quarters of the surface of the spherical earth. These ideas had no scientific or historical basis, and there were plenty of non-Christian writers who rejected them (e.g., Lucretius, Plutarch, and Lucian) or ignored them—as in the case of professional geographers such as the second-century Alexandrian Ptolemy, who concentrated his efforts in describing the known world: the Euro-Asian-African landmass or "oikoumene" (Randles 1994).

The earliest Christian mention of the antipodeans by Clement in the late first century seems to have accepted their existence. Later, when Augustine famously denied the existence of antipodeans, he did so not in association with a flat earth, as previously argued by Lactantius and later by Cosmas. Instead, Augustine rejected antipodeans on the basis of the lack of historical evidence, the speculative nature of the "symmetrical" argumentation for the antipodeans, and only later on the theological threat of having humans that were not descended from Adam or Noah. The debate continued as Medieval Christian scholars pondered over the symmetrical view of the four land masses transmitted by late-antiquity authors such as the Neoplatonists Macrobius and Capella (fourth–fifth centuries). The issue was resolved on empirical grounds during the age of exploration by Portuguese and Spanish seafarers (fifteenth–sixteenth centuries), proving that there were continents and inhabitants in all regions of a spherical earth, although not arranged in the symmetrical way Crates had expected (de Felipe 2012).

Once again, we see that a contentious issue had theologians and philosopher/scientists on both sides. Once again, we see that the issue was eventually resolved not by careful arguments but by empirical evidence—in this instance, doing the experiment of sailing around the world

## The Modern Age: Galileo and the Inquisition

Again, it is overly simplistic to speak of the famous case of Galileo versus the Inquisition as a conflict between science and faith. In fact, in the 1616 condemnation of Copernicanism, all three of the condemned books were written by avowed Christians (Nicolaus Copernicus, Diego de Zuñiga, and Paolo Foscarini). Moreover, the publication of Copernicus's *De Revolutionibus* (Copernicus 1543) had been urged by several friends of the author, all clerics: Bishop Paul von Middelburg, future bishop Tiedemann Giese, and Cardinal Nikolaus von Schoenberg—and the publication was dedicated to Pope Paul III. Later, in the 1633 trial of Galileo, his judges rightly considered themselves supported by the mainstream science of their age and the previous millennia (McMullin 2009). Galileo's support came mainly from theologians and church leaders, including disciples of the Benedictine mathematician Benedetto Castelli, and a helpful friend like the archbishop of Sienna Ascanio Piccolomini, who hosted Galileo for several months at his palace right after the condemnation by the Inquisition (Drake 1980).

Galileo thought that he had proven the Copernican system beyond doubt with his particular theory of the tides, which was probably his worst scientific blunder. It took another generation, and Newtonian mechanics, to discard Tycho Brahe's overcomplicated system (all other planets circled the sun while the sun circled the earth), and to establish the Copernican system beyond doubt. It is worth emphasizing that *all the people involved in this scientific controversy were Christians,* so framing the issue as "science versus faith" does not help in understanding the situation (McMullin 2005). Instead, the controversy pitted Christians against Christians and scientists against scientists. Theological acumen was not useful in this situation; only further scientific research could resolve the issue. Perhaps it could be seen as a foretaste of some current debates among Christians—for example, dualism versus monism—where insights from advances in neuroscience have the potential to help resolve some of these theological issues (Green 2008).

## THE CONTEMPORARY AGE:
## DARWIN AND CHRISTIANITY

It is popularly assumed that the only response from Christians to Darwin's *On the Origin of Species* (Darwin 1859) was bitter and vicious opposition based on theological prejudices. However, detailed study of the contemporary reactions shows us at least three important and often overlooked aspects that support the case we are making here. First, notable scientists Adam Sedgwick, Charles Lyell, George Mivart, Jean Louis Agassiz, and Richard Owen, although Christians themselves, opposed Darwin on real scientific grounds rather than from arguments based on theology.

Second, Christian responses to evolution were not always negative. As examples, we can point to Charles Babbage (1837), who proposed a sort of evolutionary theory long before Darwin, as well as to the writings of Charles Kingsley (1859) and Baden Powell (1860). John Henslow is also worth mentioning, as he served as Darwin's mentor, defending Darwin during a public lecture in May 1860 after Sedgwick criticized Darwin (Barlow 1967). As with the previous two examples discussed in this section, this episode was not simply a battle between theologians and scientists. Contrary to the popularized account, scientific biologists did not immediately accept Darwin's theory.

Once again there were Christians on both sides of the heated debates.

Only later would the issues be resolved by more scientific research along-side fresh biblical scholarship pointing to the necessary reappraisal of what the early chapters of the book of Genesis were and were not intended to convey (Walton 2009). Megaphone diplomacy had to be replaced by shared participation in *both* the scientific enterprise *and* a fuller awareness of advances in biblical scholarship.

Third, Darwin himself did not show an aggressive anti-Christian position, even though he had abandoned his Christian faith years before 1859. During this period of controversy, Darwin was a deist, believing in a Creator that had ordered the world by laws that were discoverable by scientific inquiry. Near the end of his life, when Darwin considered himself an agnostic, he still dismissed the inevitability of a science-and-faith conflict over evolution: "It seems to me absurd to doubt that a man may be an ardent Theist and an evolutionist" (Darwin 1879, 1).

## The Reemergence of Conflict between Psychology and Religion

Christians studying psychology in North America could be forgiven for thinking that they must be continually on the alert to resist challenges to their basic beliefs, which will inevitably arise through their psychology courses. Indeed, the need for this alertness may well be reinforced by reading books such as the two edited volumes mentioned in the preceding chapter. In those books, students would be told that there is an ongoing conflict between what psychologists are discovering and what their Christian faith encourages them to believe.

By the mid-1980s, the conflict motif was already well established in discussions of psychology and religion. For example, a book edited by Stanton Jones titled *Psychology and the Christian Faith* shows clear evidence of the conflict motif from the opening pages. In the preface, Jones wrote, "But I was constantly troubled by the apparent and real *conflicts* between what I studied at the University and what I believed" (S. Jones 1986, 11, emphasis added). A few pages later, referring to those he calls "the limiters of science," Jones claimed that their views do not help us to work out "the real *conflicts* that can emerge between psychological and religious views, such as the *conflict* between the views of conversion as a mechanistic change in attitude on the one hand and as a responsible act of a repentant human being on the other" (S. Jones 1986, 19–20, emphasis added).

Similarly, *Psychology and Christianity: Five Views* (E. Johnson 2010) contains frequent references to *conflict, battles, hostility,* and *collision of views.* The conflict motif is clearly alive and well. On the first page of the preface, Eric Johnson describes the struggle to define the role of religion in modern life as a "conceptual and political *battle*" or "culture *war.*" He asks, "What has led to this particular *conflict?*" and claims that psychology is "downright *hostile* to religion" (E. Johnson 2010, 7, emphasis added). This language certainly frames the psychology–religion relationship as a *conflict* between opposing forces.

Some books are more balanced and comprehensive in their treatment of the relationship between psychology and religion. A comprehensive book by Paul Moes and Donald Tellinghuisen titled *Exploring Psychology and Christian Faith* (Moes and Tellinghuisen 2014) largely manages to avoid the conflict motif. Nevertheless, the opening dedication page still reveals the persistence—perhaps subliminal—of the warfare motif. They wrote, "This book is dedicated . . . to all students exploring the *reconciliation* of faith and psychology." But "reconciliation" suggests that there is a conflict or battle going on with two sides that need to be reconciled.

Even an excellent book by Everett Worthington with the commendable subtitle *What Christians Can Learn from Psychological Science* sadly has as its main title *Coming to Peace with Psychology*—an expression that hints at warfare and conflict. Despite this apparent concession to the conflict motif, Worthington quickly corrects this impression. On the opening page of the introduction he writes:

> Some people believe that *science is at war with religion* and no civil dialogue can occur. I disagree. . . . In this book I will

claim that we can know people better, and even know God better, by heeding psychological science. In fact, I suggest that the relationship between psychological science and Christianity is more like an emerging marriage than either war or a mere dialogue. (Worthington 2010, 11, emphasis added)

In the section headed "The Relationship of Psychological Science to Theology," Worthington notes that two of the most frequent ways of describing this relationship have been in terms of either a "filter" model (scripture is the standard or filter for evaluating the truth of psychology) or a "perspectival" model (science and religion operate at different levels, both addressing the same aspects of reality from different perspectives). Worthington attempts to "develop a way of looking at psychological science as being in relationship to Christianity that will serve as an alternative to filter models and to perspectivalist models—a relational model." Worthington's relational model seems very similar to the theme of this book, namely, that we should seek to identify mutual insights and enrichments from the relationship of psychological science and Christian faith. Indeed, Worthington's book is filled with examples of how, in his own words, "I proposed several ways psychological science could add ideas to theology" (Worthington 2010, 214), much like the enrichments we are proposing here.

We do not deny that real conflicts between psychology and Christian faith do arise from time to time. This was recognized more than fifty years ago in an important volume on the relationship between psychology and Christian faith that recent writers in this field have largely ignored. The book was titled *What Then Is Man?* and was edited by one of the leading psychologists in America at that time, Paul Meehl. It was the result of a sustained consultation between Christian psychologists, theologians, and pastors all within the Lutheran tradition of the Christian church in the United States. They were very careful to avoid concentrating on the conflict motif, but helpfully and realistically portrayed the tension, as indicated in these words:

We have discussed some of the potentially removable sources of conflict between secular psychology and the cognitive components of Christianity. Persistent, rigorous, and creative efforts along the indicated lines will do much to improve the relation between the disciplines. It must be frankly admitted, however, that even if the factors of mutual ignorance, semantic confusion, dogmatism about open questions, incomplete empirical evidence, and philosophically unclear concepts were wholly eliminated, certain divergences would exist between the attitude and belief systems of most secular psychologists and the Christian orientation. (Meehl et al. 1958, 162)

## Challenging the Conflict-Retreat Model

Notwithstanding such balanced assessments, repeated statements about conflict and warfare by Christian psychologists look strangely out of tune with the stories told by the expert historians of science. Many books and journal articles on the history of the relations between science and religion appeared in the second half of the twentieth century and the early part of twenty-first century, and these analyses largely demolished the myth of persistent, widespread conflict in science and faith relations.

It is difficult to imagine how scholars like Eric Johnson (E. Johnson 2010) could have overlooked articles such as one that appeared in 1989 in the new journal *Science and Christian Belief.* The first article in the first issue of the journal was by Colin Russell, himself a chemist and historian of science and faith, with the title "The Conflict Metaphor and Its Social Origins." In the opening pages of this article, Colin Russell writes:

The subject of this paper is a generalization so obvious, so familiar, so pervasive, and so respectable that even to contemplate its refutation might be held to be folly, if not sacrilege. It is unlike many other myths in that it is a generalization having almost the function of a Kuhnian paradigm, a conceptual framework into which all manner of isolated and curious incidents may be easily fitted. Yet the cumulative effect of twenty years' historical scholarship has been to demonstrate its mythical character. *We are concerned with the simple thesis that science and the Christian religion are at war, and have been so for a long time.* This is not an assertion in the *philosophy* of science but rather in the *history* of science, alleging what actually happened in the past and continues to the present day. (Russell 1989, 3–4, emphasis added):

Russell later continues,

This view of the past has been variously described as a "warfare model," a "conflict thesis," and a "military metaphor." It is the intention of the present paper to suggest, not that Christian and scientific doctrines have never been at odds, but that to portray them as persistently in conflict is not only historically inaccurate or actually a caricature so grotesque that what needs to be explained is how it could possibly have achieved any degree of respectability. (Russell 1989, 5)

Recent books by authors such as Ronald Numbers (Numbers 2009), John Brooke (Brooke 1991), and Peter Harrison (Harrison 2010) document the varied interactions between science and faith over many centuries. They provide a wealth of lessons to be learned about how properly and constructively to relate knowledge from the scientific enterprise with knowledge from other domains of knowledge, including religion. Harrison's book argues that there are no easy answers to understanding the complexities of the issues involved. However, so important are the potential lessons to be learned from history that it is crucial that any new science, such as psychology, should be aware of the rich and complex history of these past interactions. Specifically, psychology needs to learn from the past—looking for examples to follow or errors to avoid—as it seeks to work out its relations with religion.

Is there something fundamentally different about psychology as a science? If not, then why have psychologists not learned these lessons? Why has the conflict motif persisted? Of course, it is possible to select historical examples on which to build a narrative in which science drives religion into permanent retreat. However, a closer examination of science-and-religion interactions spread over two millennia reveals that such a "conflict-retreat" portrayal of science-religion relations tells only part of a story that is, as Harrison has documented, much more complex. Christian institutions and Christian scholars often directly supported and advanced the cause of science. At other times, Christianity, under the guise of a foe, did the work of a friend for science. With such examples in mind, we propose that there is a more constructive way forward.

Discussions of the relations between science and religion typically refer to Ian Barbour's book *Where Science Meets Religion*. They note Barbour's four-way classification system of the possible relations between science and religion: *conflict, independence, dialogue,* and *integration*. Barbour's own sympathies are with *dialogue* and *integration*. Thus, he writes at the end of his book, "In summary, I believe that dialogue and integration are more promising ways to bring scientific and religious insights together than either conflict or independence" (Barbour 2000, 179). It is interesting that he refers here to *insights*—a major theme of this book. He is, however, clearly aware of the dangers of some forms of integration that look suspiciously like concordism. He argues that problems will arise "if either scientific or religious ideas are distorted to fit a *preconceived synthesis* that claims to encompass all reality" (Barbour 2000, 45, emphasis added).

Alister McGrath, in his 2014 Inaugural Lecture when taking up the Andreas Idreos Professorship of Science and Religion at Oxford University, asked, "So how did we get there, when historical evidence makes such a narrative [of conflict] so problematic?" McGrath continued, "Yet to those in the know, this 'science versus religion' narrative is stale, outdated and largely discredited. It is sustained not by the weight of evidence, but by endless uncritical repetition, which studiously avoids the new scholarship which has undermined its credibility" (McGrath 2015, 7). Later McGrath said that "we should challenge the dominant narrative of our time—the outdated 'conflict narrative,' sustained more by uncritical repetition than by historical evidence—and replace it with a narrative of enrichment" (McGrath 2015, 12). It is noteworthy that McGrath here uses the idea of *enrichment* as a useful way to move forward, echoing the theme of this book.

In a seminal paper, David Lindberg and Ronald Numbers pointed to the need to contest the traditional examples, the causes célèbres, of the conflict model throughout history:

> Recent scholarship, however, has shown the "warfare" thesis to be a gross distortion—as this paper attempts to reveal, employing illustrations from the patristic and medieval periods and from the Copernican and Darwinian debates. (Lindberg and Numbers 1987, 140)

Although Lindberg and Numbers do refer to the patristic and medieval periods, the thrust of their paper is devoted to the Copernican and Darwinian debates. Apart from debunking many false pseudohistorical details in the literature on conflicts, the main straightforward way to confront such biased historical reconstruction is to realize that these debates were hardly "science vs. religion." As a host of historians have shown, in each of these confrontations there were Christians and frequently even scientists (as well as persons who combined both trainings) on both sides of the argument, as illustrated in the particular case of Galileo:

> The Galileo affair . . . was not a matter of Christianity waging war on science. All of the participants called themselves

Christians, and all acknowledged biblical authority. This was a struggle between opposing theories of biblical interpretation: a conservative theory issuing from the Council of Trent versus Galileo's more liberal alternative, both well precedented in the history of the church. (Lindberg and Numbers 1987, 145)

So, we see that some authors, by selecting particular historical examples, were able to weave a narrative about

- A persistent, widespread *conflict* between science and religion, which leads inevitably to
- A continuous *retreat* from the positions of Christians who "got it wrong" on scientific principles.

A different set of historical examples, however, shows that this has not always been the case. The founding father of the big bang cosmology, the priest and scientist Georges Lemaître, had "no conflict to reconcile" between his faith and his scientific work (Aikman 1933, 18). In cases like this, there is no warfare, no battle lines are drawn, and no retreat is necessary. Other similar examples show that the history of science and faith relations is more complex than the conflict model assumes.

Furthermore, as de Felipe and Jeeves (2017) have documented, we believe there is a third set of examples where a rather unexpected interpretation can be made. Interestingly, in these cases the conflicts between science and faith have not arisen from a desire by Christians to fight against science, but rather the opposite—from a desire to find some sort of concordism between science and faith. Indeed, most past concordisms between science and faith have ended up producing unnecessary conflicts when science has moved on and left faith exposed, hanging on to a "god of the gaps" when there is no longer a gap to fill. A compelling example is the realization that, to a significant extent, Galileo's troubles were the consequence of centuries of an unhelpful synthesis or integration of Christian theology with the teachings of Aristotle. Thus, the conflicts of one era are frequently the catastrophic consequences of concordism from a previous era.

The take-home message from this brief and highly selective overview is that there is a robust historical basis for rejecting the *conflict-retreat model*, and that this evidence is persuasive to leading historians of science, as indicated in this conclusion from Peter Harrison:

> When examined closely, however, the historical record simply does not bear out this model of enduring warfare. For a start, study of the historical relations between science and religion does not reveal any simple pattern at all. In so far as there is any general trend, it is that for much of the time religion has facilitated scientific endeavor and has done so in various ways. (Harrison 2010, 4)

We suggest that it is time to move beyond the *conflict-retreat* paradigm, not only through a detailed clarification of these specific episodes, but also by broadening the discussion to consider other historical episodes that illustrate an even more complex and more representative account of science-faith relations.

## The Relevance of Advances in Biblical Scholarship to Discussions of Science and Faith

Proponents of the conflict motif also tend to overlook the fact that biblical interpretation has changed across the centuries, and advances in biblical scholarship have actually diminished some of the apparent conflicts between science and theology. An early example of a "science-friendly" scholar was Jonathan Edwards, one of North America's greatest theologians. Edwards had good friends who were distinguished scientists and Fellows of the Royal Society in London. His correspondence with those scientists shows that he was very interested in scientific discoveries that might have implications for traditional interpretations of the scripture passages that refer to God's creation. In a biography of Jonathan Edwards, George Marsden wrote, "Edwards regarded Scripture alone as truly authoritative, so earlier interpreters could be revised. The project of understanding Scripture's true meaning was an ongoing progressive enterprise to which Edwards hoped to contribute" (Marsden 2003, 474). Edwards's emphasis on the dynamic nature of scripture interpretation is relevant to the case we are trying to build. It is scripture itself that is authoritative, not the interpretation given by a particular group of Christians at a particular time in history. Going one step further, as N. T. Wright does, we could say that it is God who is authoritative and who speaks through scripture (N. T. Wright 2011).

Many Christians see a fundamental difference between a scientific understanding of the world that changes with each new scientific discovery and a theological understanding that is static and unchanging. We believe that this view is not realistic. As John Walton, a leading biblical scholar in the tradition of Jonathan Edwards, wrote:

> Though the Bible itself does not change, we realize that our interpretation of Scripture is much more dynamic and the resulting shape of theology consequently subject to constant reassessment (more on the perimeter than in the core). Two millennia of church history have witnessed some dramatic differences in hermeneutics, some deeply ingrained theological controversies (some options cast off as heretical, some bringing major splits and some being retained side-by-side) and some substantial disagreements about the interpretation of particular passages.... *New insights and new information can emerge at any time.* Several hundred years ago, renewed access to the original languages had significant impact on biblical interpretation. In recent decades, the availability of documents from the ancient world has provided a remarkable resource for our reading of the biblical text. We dare not neglect these tools when they can contribute so

[ 37 ]

significantly to our interpretation. (Walton 2015, 11–12, emphasis added)

A similar case was made by Peter Enns, another leading biblical scholar, who urges us to keep three ideas in mind when approaching scripture:

> Our knowledge of the cultures that surrounded ancient Israel greatly affects how we now understand the Old Testament—not only here and there but also what the Old Testament as a whole is designed to do.
>
> Because scripture is a collection of discrete writings from widely diverse times and places and written for diverse purposes, the significant theological diversity of scripture we find there should hardly be a surprise.
>
> How the New Testament authors interpret the Old Testament reflects the Jewish thought world of the time and that accounts for their creative engagement of the Old Testament. It also helps Christians today understand how the New Testament authors brought together the Israel story and the Gospel. (Enns 2012, introduction, xi)

Thus, biblical scholars remind us that proper interpretation requires careful study of the ancient biblical texts in the light of new archaeological evidence, new textual studies, and new social-cultural studies of life in biblical times. For example, in the study of Old Testament passages describing the events of creation, it is essential to understand the structure of Hebrew poetry and its use of powerful cultural images to convey a message. Similarly, New Testament scholars cannot understand the message of the parables of Jesus without placing them in their cultural and historical context. *God's message always comes in a form that is incarnated in a specific ethnic, cultural, historical, and linguistic context.* In the Bible, we have a written record of that revelation, wrapped in literary forms that were in common use during biblical times. Of special interest to psychologists are those passages of scripture that contain insights into our mysterious human nature—using terms such as *mind, heart, soul, spirit,* and *flesh.* We can so easily read them as if they address our post-Enlightenment scientific concerns, when in fact they do not. A notable recent example of how biblical scholarship can illuminate our understanding of these passages is a book by Joel Green titled *Body, Soul, and Human Life: The Nature of Humanity in the Bible* (Green 2008).

## Applying Lessons from History

The contributing authors to one recent volume (E. Johnson 2010) engaged in a healthy debate, expressing diverse views about how to relate psychology and the tenets of Christian faith. From our point of view, several of the chapters offered thinly veiled attempts at concordisms, revealing presuppositions about both the purpose of scripture and the nature of the scientific enterprise. For example, one essay (Roberts and Watson 2010) described a "Christian psychology view." In reading that chapter, it is difficult to avoid the conclusion that the authors believe that *all that we need to discover about psychology is somehow and somewhere buried within scripture.* To make progress in discussions of the relation of psychological science and Christian faith, we need to bring to light these presuppositions about the nature and purpose of scripture. The chapter by Roberts and Watson evoked this response from David Myers, another contributor to the book:

> I concur with Roberts and Watson's esteem for the rich insights of the ancient philosophers, of Jesus and of theologians from Augustine to Kierkegaard, but I do so without conflating them all as psychology. By nearly all definitions, psychology is today's science of behavior and mental processes. So, while Roberts and Watson may dismiss today's psychology—that of national psychology organizations, of leading universities, of

psychology exams and of psychology texts—as mere "establishment psychology" and instead advocate Christians to gather together on their own to create a different psychology, that's a bit like dismissing the American Medical Association or the National Institutes of Health or university medical schools as offering "establishment medicine." So they are, but Christians are ill-advised to scorn their achievements and to absent themselves from medicine's practice and leadership. (Myers 2010b, 179–80)

In this same volume, Stanton Jones advocated an "integration view" of the relation between psychology and Christianity (S. Jones 2010). We examine this widely used notion of *integration* in later chapters, as not everyone writing about integration is as careful as Jones in making clear what it does and does not involve. However, on the specific issue of the use of scripture to settle arguments, we want to point to the response of David Myers to the chapter by Jones. Although finding much with which he can agree, Myers suggests that Jones overstates the extent to which scriptural truths are "as stable and enduring—or more so—than any empirical generalization," given that new work by biblical scholars and theologians can lead to significant revisions in our understanding of biblical passages (Myers 2010c, 130).

This last point about the impact of new biblical scholarship was also highlighted by Everett Worthington:

> The validity of theology is not above suspicion. It roots its understanding of texts to a particular historical context. Effort is made to understand the historical context, word usage and culture at the time the text was written with as much accuracy as possible. Then theology projects forward from the historical context of the present. Biblical interpretations are thus based on our understanding of context and on what can and cannot be generalized to the present. The passages in Paul's

letter to the Corinthians on women speaking in church, head coverings, slavery and other issues come to mind. Theologians either must dismiss some advice as irrelevant or try to discern the principle relevant for today. If our understanding of the historical context changes, then our interpretation of Scripture can change. (Worthington 2010, 203)

## Summing Up

This chapter has focused on the origins of the conflict model and the historical examples that have been used to promote that model. We attempted to show that these examples do not actually support the conflict model. Instead, a strong case can be made that *the idea of persistent, inevitable conflict between science and religion is a myth*. We also argued that new developments in biblical scholarship have reduced some of the apparent conflicts between theology and science.

We conclude by restating the need to be open-minded but not empty-minded about what these developments can teach us. We must reflect on what implications the advances in biblical scholarship may have for some of our earlier formulations of what scripture does and does not teach about humans and our world. We must also reflect on what implications new scientific discoveries, especially in neuroscience and evolutionary psychology, may have for our earlier theories about human nature. These issues are our focus in later chapters.

CHAPTER 3

# From Conflict to Concordism

## Alternatives to the Conflict Model

Historians of science have carefully traced the relationship between science and Christianity across the past two centuries, as well as the relationship between Christianity and natural philosophy in the centuries before modern science developed (e.g., Brooke 1991). Two recurring themes have emerged from this study:

- Repeated claims and counterclaims about the existence of *conflicts* between science and faith—what Peter Harrison calls the "conflict myth" (Harrison 2015).
- Repeated attempts to develop *concordisms* between the scientific consensus about a particular issue at that point in history and the dominant interpretation of relevant biblical passages in that same period (Davis 2006).

We have seen that conflicts easily arise when statements about the physical environment or human nature that come from two distinct language domains are put alongside each other as if they are competitors. Is there a more appropriate way to relate claims about a set of events from the perspective of science with claims about those events from the perspective of theology? One example (mentioned in chapter 2) is provided by the founding father of big bang cosmology, the priest and scientist Georges Lemaître. Recognizing that the claims and explanations coming from within science and from within theology are based on different assumptions and serve different purposes, Lemaître argued there is "no conflict to reconcile" between his faith and his scientific work (Aikman 1933, 18).

In recent years, some psychologists who see no necessary conflict between their faith and their science have worked toward the "integration" of science and faith. That is an understandable position to be taken by a Christian who believes that the same God "who upholds all things at all times by the word of his power" (Hebrews 1:3) is also the God whose world we are investigating as scientists. It would seem obvious that we should try to "integrate" the accounts given within scripture and the accounts we produce as scientists. Indeed, one of us (MAJ) has, in earlier years, made reference to integration as a worthy goal (see Jeeves [1997] 2006), without implying any kind of improper intermixing of the statements of science and the words of scripture. In retrospect, mindful now of the pitfalls associated with loose and ill-defined uses of the word "integration," it is possible to see that different terminology would have been preferable. Given these second thoughts, in a later chapter we shall put the uses of the word "integration" under the microscope to see what we can learn from both its proper use and its misuse.

## THE ALLURE OF CONCORDISM

### Understanding Concordism

Scattered across the history of science are episodes in which a new scientific development demolishes a previous alliance between science and religion, generating a new—and unnecessary—conflict. Some religious people interpreted each episode as an opportunity to construct a new concordism to harmonize their theological beliefs and the latest scientific theories. But that merely started the clock ticking again in a new countdown until the next presumed conflict emerged.

In a talk given at a Christians in Science Conference in Oxford, Pablo de Felipe recounted the long history of concordism in Christian thought. He said:

> Concordism was not invented by Philoponus or Basil. It was a constant in early church history. And in some ways, it was a positive move towards science. Far from trying to destroy science, the attitude of most early Christian leaders was to find integration. . . . The way to achieve so was concordism. If the God of the Bible is the creator of the cosmos, the information conveyed by his Word revealed in the Bible has to agree with his creation activity as found in the cosmos. (de Felipe 2017)

As de Felipe noted, the logic behind concordism flows from the belief that the same God who is the source of revealed Christianity is also the Creator of the world. Across the centuries, this view was often referred to using the "metaphor of two books." Proponents held that the truth of our existence is contained both (and equally) within the "book of scripture" and the "book of nature" (Harrison 2006). This perspective assumed that Christianity and science must ultimately be in concordance, as two truths cannot contradict one another. The idea of a unified truth that merges science and faith has been an important guiding principle in Christian theology. Consequently, Christians aligned themselves and their theology with the great intellectual systems initiated by Plato, Aristotle, and other ancient philosophers, culminating in St. Thomas Aquinas's ambitious synthesis of Christian theology and Aristotelianism (Thomas 1912).

## *Accommodation Theory:*
## *Protecting Concordisms from Contradictions*

However, problems arose when a clear contradiction between science and theology appeared to emerge—especially when the information derived from the natural world didn't match the explanations derived from scripture. In the ancient church, the clearest example was the problem of the shape of the earth. One possible solution to the conflict was to reject science, or rather to create an alternative "domesticated" science, as the sixth-century Alexandrian monk Cosmas Indicopleustes did when he proposed a scripture-based theory of a flat, rectangular earth (Cosmas 1897).

A different solution, already discussed during the patristic era and promoted in Western Christianity chiefly due to the influence of Augustine, was the so-called accommodation principle. According to de Felipe:

> Early Christians devised a solution, the accommodation principle: if biblical texts were written in a particular place and time, it should be expected that the understanding of nature in those texts should be according to the knowledge at that place and time, as divine revelation should not be expected to be concerned with those issues. We can see a glimpse of that in Philoponus's defense of Moses as writing a book not to teach astronomy, but to lead his people to God. (de Felipe 2017)

According to this view, the Holy Spirit accommodated to the cultural level of the ancient recipients of scripture. Therefore, when reading the Bible, one should not be surprised to encounter incorrect or incomplete statements about scientific topics.

John Calvin, a leading proponent of accommodation, argued that "the Bible is primarily concerned with the knowledge of Jesus Christ. It is not a scientific textbook. God has to come down to our level to reveal himself, but has need to accommodate his language to meet our limited abilities and understandings" (Hawkes 2012, 9). In his commentary on Genesis, Calvin applied the accommodation principle to questions about the shape and movement of the earth:

> For, to my mind, this is a certain principle, that nothing is here treated of but the visible form of the world. He who would learn astronomy, and other recondite arts, let him go elsewhere. (Calvin 1847, vol. 1, 1.6, 79)

Moses wrote in a popular style things which, without instruction, all ordinary persons, endued with common sense, are able to understand; but astronomers investigate with great labor whatever the sagacity of the human mind can comprehend. (Calvin 1847, vol. 1, 1.16, 86)

Calvin's views were shared by several leading astronomers during the Copernican crisis. As one example, Danish astronomer Tycho Brahe described his position in a 1589 letter to astronomer Christoph Rothmann:

For although they [the sacred writings] adjusted themselves to the common method of understanding . . . yet let it be far from us to think of them as speaking in such a common manner that we do not believe them to be speaking truth. Thus Moses, even if he does not refer to the deep things of astronomy when treating the creation of the world in the first chapter of Genesis, because he is writing for the common people, nevertheless he does introduce that which our astronomers can concede. . . . Although they [the prophets] did not treat of physics by profession, indeed this was not the nature of their gift, nonetheless they mixed many physical propositions in with their prophecies. (Brahe 1913–29, vol. 6, 177–78)

Rothman himself went even further, claiming that the Bible has absolutely nothing to say about scientific astronomy, because "God speaks, accommodating Himself to the capacity of the Hebrews" (Granada 2008, 571).

### Accommodation Leads Back to Concordism

A radical form of accommodation theory allowed some Christians to avoid choosing between theological explanations and scientific explanations—considering science and faith as two completely different areas of inquiry ("two books") that could coexist without needing concordism. This position remains viable in modern times, as demonstrated in the quotation from Georges Lemaître above, and by the widely publicized "nonoverlapping magisteria" approach proposed by Steven Jay Gould (Gould 1997).

However, accommodation theory often ended up leading back to concordism again by considering that, although the Bible could be interpreted in simplistic ways that matched the "folk science" understanding of the common people, it could (and even should) be interpreted in sophisticated ways to be in concordance with "true natural philosophy" or "accurate science." To achieve such concordism, its defenders chose either to distort the Bible, distort the scientific findings, or both.

Once a concordism had been constructed—such as the harmonization of Ptolemy's earth-centric cosmology and biblical descriptions of the heavens—Christians had a very hard time abandoning the concordism, even after science had moved on.

Concordist accommodation is still very common among some Christians today. Unfortunately for concordists, over time, as science moved on to new understandings, *religious leaders were often left alone defending an old, unsupportable concordism between their theology and an out-of-date science.* As Brooke has described it:

> A recurrent problem was that the harmonization could work too well—creating an embarrassing situation when the science moved on. . . . The more successful the harmonization, the more trouble it caused when the science moved on. (Brooke 2009, 94)

One special case of concordism is the "god-of-the-gaps" view, where science and faith are not really harmonized, but merely juxtaposed in such a way that religion fits only into the spaces where scientific knowledge is incomplete. The problem with this approach is that, in due time, each gap is filled with new scientific findings, making God increasingly superfluous. In this way, the supposed scientific support for a theological principle becomes transformed into an embarrassing obstacle to faith. When religious people hold fast to out-of-date science, they make things worse for other believers. That leads to a paradoxical conclusion: many of the apparent science and faith conflicts are the result of previous science and faith concordisms.

The ideal of a perpetual alliance with a "true" and "static" science is in itself misguided, because the nature of science demands continuous progress to better understand the world. This point was reinforced by Nobel laureate Venki Ramakrishnan, the current president of the Royal Society, who wrote:

> The court of science never passes a final judgment, but constantly re-evaluates the evidence to arrive at our current understanding. . . . The evidence will win out in the end. Science . . . is not perfect, but science is still our best bet for understanding the world around us and for improving our lives. (Ramakrishnan 2016)

## Failed Concordisms

As science progresses, forced harmonizations typically turn into failed concordisms. The well-documented histories of science and faith already referred to several times (Brooke 1991; Numbers 2009; Harrison 2015) show that those resisting new scientific ideas often did so using not only religious grounds but also scientific ones, derived from the established science of their time that had formerly been considered to be successfully harmonized with religion. These episodes then became classical examples of failed concordisms. For example, Cosmas used ancient Greek geographers to support his flat-earth theory against the "modern" scientific view of a spherical earth (Cosmas 1897). Augustine based his rejection of the antipodeans on classical geography that dismissed the possibility of communication between the Northern and the

Southern Hemispheres (de Felipe 2012). Copernicanism was condemned on philosophical and scientific grounds as well as on religious grounds (McMullin 2009). Critics of Darwin not only used theological arguments but also were confident that Darwinism was not sufficiently grounded in the scientific evidence (Brooke 1991). Of course, in these cases there was a religious agenda to defend, but the defenders did have real scientific support for their views at that point in history (Numbers 2009).

In the previous chapter on the ubiquity of the conflict model in the history of science and faith relations, we argued that, although the popularizers of the conflict model have habitually portrayed Christians as conspiring against science, the truth is frequently the opposite. The prestige, first of the natural philosophers and then of the scientists, has been so great that religious leaders often tried to mobilize their support—at times at the price of making intellectual concessions—to achieve a concordism. But now we ask, were these concessions too great a price to pay? All too often, the carefully constructed concordisms of one age sowed the seeds of unnecessary conflicts in subsequent periods. This has already happened in psychology's short history.

Within the topic of psychology and religion, there are two examples of failed concordisms that are especially noteworthy. They provide lessons to be learned and applied to contemporary projects that relate psychological science and Christian faith. The first is *the attempt to harmonize biblical statements about human nature with a dualistic anthropology derived from Plato.* The second is *the wide acceptance of phrenology by theologians, seeking a concordance between Christian doctrine and what was then the "modern" science of brain and behavior.* We explore each of these historical examples and then consider whether additional failed concordisms have emerged in the past few decades from clinical and counseling psychology.

### Dualistic Anthropology in the Early Church and Beyond

Contemporary biblical scholarship has challenged the dualistic view of the soul as separate from the body, urging us to return to a more biblical model of personhood that embraces the psychobiological unity of a human being. However, some form of dualistic anthropology seems to have been widely considered as self-evident by theologians from the second century onward. This perspective of a nonmaterial soul controlling a physical body became part of mainstream Christian theology across many centuries. It is interesting that the dualistic view, not coming originally from the Bible, was nevertheless readily incorporated by the early Christians into their view of the person. Why did this happen? Apparently, theologians deferred to the higher "scientific" status of the Greek philosophers. As John Cooper stated, "Given its authoritative position, the Platonic body-soul distinction was uncritically accepted for more than a millennium" (Cooper 2000, 26).

One of the reasons that dualism persisted for so many centuries is the perception that dualistic anthropology is necessary to defend the existence of the soul. The importance of the concept of the soul for theology was described by New Testament scholar Joel Green:

> Theologians ancient and contemporary have found in an anthropology of body–soul dualism either the necessary supposition or corollary of a number of theological loci, including creation in the divine image, a theology of free will and moral responsibility, hope of life-after-death, and Christian ethics. . . . As early as the second century of the Christian era it was nonetheless clear to most theologians, as the *Epistle to Diognetus* (§6) puts it, that "the soul lives in the body, but it does not belong to the body"; indeed, "the soul, which is invisible, is put under guard in the visible body" and "the soul is imprisoned in the body, but it sustains the body." (Green 2008, 18–19)

These dualistic views quickly became widely held presuppositions in the early church. "Without the soul, we are nothing," wrote Tertullian (*On the Flesh of Christ* 12). Lactantius defended dualism by claiming that the body, formed from the earth, is solid and mortal, "made up of a ponderous and corruptible element," but the soul "received its origin from the Spirit of God, which is eternal" (*Divine Institutes* 12).

Thus, we see the enduring, pervasive power and academic respectability of Platonic dualism continuing to be felt long after the death of Plato in 347 BCE. After the early Christian church left the Jewish context of its early development during the first century, it found itself immersed in a Neoplatonic milieu. The admiration of Plato in the early church was so great that some theologians believed that Plato was to some degree inspired by the Holy Ghost, and others believed that the dualist philosophers had stolen their ideas from the Bible (see Augustine, *City of God* VIII.11). Paradoxically, they did not realize that it was in fact the other way around: theologians had adjusted Christian teachings in order to conform to Platonism.

The influence of Platonic dualism became solidified in Christian theology after Augustine blended it with biblical teachings in an early example of concordism. Although Augustine had profound differences with Tertullian, both shared a dualistic approach that incorporated what they considered the best science available at the time. According to Robert Brennan,

> Augustine follows Tertullian at one point where Tertullian reverses his normal rhetoric to base his understanding of anthropology and in particular of the origin and nature of the soul upon contemporary *scientia* rather than his usual practice of beginning with Scripture. (Brennan 2015a, 85)
>
> Augustine engaged with the best of contemporary philosophy regarding nature and medicine. . . . Augustine, like Tertullian, utilizes and trusts references to contemporary medical terminology, albeit with further centuries of scholarship behind him. (Brennan 2015b, 76)

Platonic anthropology remained deeply entrenched in Western culture and Christian tradition. When the thirteenth-century theologian–

philosopher Thomas Aquinas adopted Aristotle as his philosophical guide in building up a new concordism to replace the old Platonic one, dualism was retained—in clear opposition to the monistic-materialistic view of Aristotle, who had challenged the dualism of his teacher Plato. On this contradictory situation, William Hasker has written:

> Unfortunately, Thomistic dualism suffers from a fundamental incoherence. The key idea, taken from Aristotle, is that the soul is "the form of the body." On the face of it, this seems to say that the soul is a sort of pattern or structure. . . . So understood, this view would really be a kind of materialism. . . . However, this is not what the Thomistic dualists really want to say. According to them, there are certain mental activities, involving abstract reasoning, that are performed only by the soul, with no assistance from the body or brain. Clearly this would not be possible if the soul were understood merely as a pattern or structure of the body. (Hasker 2010, 94–95)

During the twentieth century, theologians began to question Platonic body–soul dualism and work toward the recovery of a more holistic "Hebrew view" of human beings. For example, Old Testament scholar Bill T. Arnold has written, "Recent studies have admitted the Hebrew Bible's purely physical perception on human personhood, acknowledging the impossibility of developing a Christian dualist anthropology on the basis of these data" (Arnold 2004, 83).

Interestingly, this shift in biblical interpretation happened around the time that discoveries in neuroscience began to emphasize the fundamental mind-brain or mind-body unity of the human person. Today, as neuroscience continues to demonstrate the essential connection between mind and brain, Christians are often criticized for supporting an anti-scientific dualistic view of humans—providing a clear example of an ancient science-driven concordism being finally exposed. In response, contemporary theologians face an almost irresistible temptation to claim a *new concordism* between today's "more accurate" science and the "more appropriate" biblical view of persons, once again using science to "prove" the truth of religion. *But scientific progress will continue, and biblical scholars will continue to surprise us with their findings and challenge us to new and deeper understandings of what God has revealed.* Scripture was never intended to be used to build scientific models of persons or any other part of creation. To do so invites adding to the long list of failed concordisms.

## *Phrenology and Neurotheology*

Throughout the medieval period, philosophers and theologians assumed that the mind was composed of distinct "faculties" or modules, with each module performing a specific mental task. Perhaps the first serious attempt to link this "faculty psychology" view of the mind to the physical brain was made by Viennese physician Franz Joseph Gall, who began his work in the 1790s by collecting more than three hundred skulls and brains. Gall proposed that the brain's cerebral cortex is made up of individual organs, each of which controls a particular mental faculty (Gall 1835b). Gall's collaborator, the German physician Johann Spurzheim, popularized the term *phrenology* to describe Gall's theory. Phrenology involved reading someone's character, personality, and abilities from the shape of the skull. It was based on the idea that the mind (and its components) was ultimately and intimately related to the physical brain. Phrenology implied some kind of localization of the functions of the brain. Enlarged areas of the brain, signified by bumps or protrusions on the skull, were thought to indicate the locations of character traits or mental abilities (Spurzheim 1840).

Contemporary neuropsychology looks back at phrenology with puzzled condescension, regarding it as a primitive effort to understand the brain-mind relation. In the context of this book, the relevant issue is the fact that some leading Christian thinkers of that era embraced the "new science" of phrenology with enthusiasm. It was a short-lived concordism that proved to be a serious mistake, not because Christianity rejected science, but because these Christians accepted uncritically a set of speculative scientific principles that were later discredited.

It is possible to distinguish between two groups of early phrenologists who held differing religious beliefs, although the dividing lines between these groups is at times fuzzy. One group, including Gall, Spurzheim, and George Combe (who founded the first phrenological society), were deists. They believed that religious beliefs should be founded on human reason and on the observation of nature. Their attempts at concordism were skewed in the direction of giving primacy to the new science, in the sense that revealed truths from scripture could not possibly conflict with observed truths from nature (Combe 1824). The second group could be labeled "Christian phrenologists," because they all subscribed to the basic tenets of Christianity and attempted to construct detailed concordisms between Christian theology and "scientific" phrenology. It is interesting to see how these two distinct groups of religious thinkers could each mobilize the new science of phrenology and harmonize it with *their* distinctive religious views.

A brief look at some of these Christian phrenologists demonstrates the way in which they made use of phrenology to support their Christian beliefs, producing concordisms that were not able to stand the test of time (see Norman and Jeeves 2010). Henry Clarke, a Scottish theologian who wrote *Christian Phrenology*, regarded phrenology as a friend and helpmate to Christianity, especially as it explained some aspects of a person's moral nature (H. Clarke 1835). His contemporary, British medical doctor Charles Cowan, wrote *Phrenology Consistent with Science and Religion* to explicitly advocate a "harmony" between scriptural Christianity and the principles of phrenology. For him, since God is the source of both nature and revelation, such harmony must necessarily exist (Cowan 1841). Another attempt at concordism came from William Scott, who wrote *The Harmony of Phrenology with Scripture* to argue that philosophy and phrenology must be considered subordinate to enlightened faith (Scott 1837). Later, William Easton wrote a book titled *The Harmony of Phrenology and Scripture on the Doctrine of the Soul*. Similar to the deist phrenologists, Easton gave greater importance to the truths of nature as revealed in phrenology research, arguing that the truths of scripture must be modified to match the findings of phrenology (Easton 1867). In North America, the most prolific of the Christian phrenologists was Orson Fowler. In his book *Religion: Natural and Revealed*, he promoted a form of phrenology that provided a scientific basis for a "purified" type of religion, claiming that spirituality is built into the organization of the brain; within the brain, humans have an organ of spirituality in the same sense that we have organs designed for vision and hearing (Fowler 1844).

A review of these champions of phrenology summarizes the situation:

> By considering the views of this small sample of nineteenth-century Christians and deists seeking to relate the nineteenth-century "brain science" of phrenology with their religious beliefs, we find a wide variety of proffered solutions. Some wanted to replace religion with science (Combe), some to purify religion (Fowler), some to find in science a friend and helpmate (Clarke), some to harmonize science and faith (Cowan and Scott). (Jeeves and Brown 2009, 36)

Apparently, those claiming to be Christian phrenologists found ways to accommodate the materialistic "brain controls mind" approach of phrenology while still maintaining their Christian faith. But it was a wasted effort. Phrenology was abandoned by brain scientists in the twentieth century as a mistaken approach to the study of the mind-brain relationship. As phrenology was forgotten, the concordisms between religion and phrenology also disappeared.

In this context, it is interesting to observe the enthusiasm with which some contemporary Christian leaders have embraced *neurotheology*—which is, in some respects, a modern-day phrenology. Are there specific areas in the brain that are involved in, or even dedicated to, religion or spiritual experiences? There is certainly some evidence pointing in that direction. For example, at a meeting of the Society for Neuroscience in the United States, the well-regarded neuroscientist Vilayanur Ramach-andran described two patients with temporal lobe epilepsy who, when they were presented with religious stimuli, showed an unusually strong galvanic skin response, which is a sign of emotional arousal (Ramachan-dran et al. 1997). It didn't take long for provocative labels such as the "God spot" to be applied to regions in the temporal lobe associated with this condition, or to an area of the parietal lobe that becomes especially active during religious meditation. For some, the temptation to enlist neuro-theology to "prove" the existence of God has become irresistible, leading to the publication of books with eye-catching titles such as *Where God Lives in the Human Brain* (Albright and Ashbrook 2001). The back cover of the book provocatively asked, "Are we 'hardwired' to seek God?" That may have helped to sell the book to those seeking scientific evidence to prove the existence of God, but it certainly does not do justice to the careful treatment of the topic within the book.

In a recent book, Peter Clarke wisely commented, "Thus neurosci-entific studies of religious experience may tell us which parts of the brain are involved, but they leave entirely open the question of whether the experiences reflect a spiritual reality" (P. Clarke 2015, 191). Lesson learned: *it is unnecessary and unwise to treat weak evidence for a "God spot" in the brain as showing "proof" of the existence of God or a new concordance between science and faith.*

### Counseling Psychology and Psychotherapy

Of all of the subfields of psychology, the areas of clinical and counseling psychology have received the most attention from Christians, perhaps because they see natural connections between psychological concepts such as "anxiety" and "depression" and theological concepts such as "sin," "evil," and "forgiveness." We have already referred several times to a book by Everett Worthington titled *Coming to Peace with Psychology: What Christians Can Learn from Psychological Science* (Worthington 2010). Worthington, who describes himself as both a counselor and a psychological scientist, provided a careful, thorough, and sympathetic account of the challenges in seeking properly to relate a biblically based Christian faith with these specialist areas of psychology. As he pointed out, it can be a demanding task for Christian counselors and psychotherapists, but their clients were demanding it:

> The Christian community cried out for explicitly Christian counseling that was centered on biblically consistent beliefs and values—traditional theology that one could count on— and on counselors who took Jesus Christ seriously, prayed in sessions, spoke openly about spiritual concerns and assigned homework centered in stable truth, the Bible. . . . These clinical psychologists became the early integrationists. (Worthington 2010, 34)

What is meant by "integrationist" here, and what does it look like in practice? According to Worthington, some of the early leaders in the field (such as Clyde Narramore) "adapted the secular psychotherapy theory in which they had received their training" (Worthington 2010, 35). In the case of Narramore, Freud's theories became the basis for integration. At this time, detailed scrutiny and testing of Freud's theories were still taking place, and Narramore could perhaps be forgiven for integrating his Christianity with psychological theories largely discredited today. Is this another example of a failed concordism?

On the views of Narramore and others, Worthington wrote, "The religious tradition was the mold, and the secular theory of psychotherapy was fit into it" (Worthington 2010, 35). We suggest that any attempt to fit a particular psychological theory into a particular religious tradition is to totally misunderstand both psychological theory and religious tradition. Worthington provided another metaphor for thinking about this type of integration when he said, "[In Christian counseling,] the secular counseling theory is essentially poured through a theological filter" (Worthington 2010, 35), which presumably would exclude any troublesome psychological details that did not fit well with the counselor's religious beliefs. Worthington listed a number of pioneers in Christian counseling, including Gary Collins, James Beck, James Dobson, and Larry Crabb, and described how they tried to build an explicitly Christian approach to counseling upon the foundation of the particular secular theory of psychotherapy in which they had been trained. *In effect, they were all trying to integrate Christian faith and psychotherapy, but the result produced yet another concordism.*

As one example, Worthington referred to a recent integration model proposed by Eric Johnson which "insists that Christian theological worldviews be put forth to understand the data of psychology." Worthington responded:

> While I believe this approach will be wonderful for journals like the *Journal of Psychology and Christianity* and the *Journal of Psychology and Theology*, which accept Christianity, I fear that his proposal will have no hearing in mainstream psychology—either clinical or psychological science. It would, I believe, be flatly rejected. And even if it were to be accepted, it would marginalize Christian psychology, taking it out of the mainstream and creating Christian psychology backwaters in which it would float inertly. (Worthington 2010, 40)

We fully agree with Worthington's assessment. Johnson's approach risks producing fuzzy concordisms that will not do justice either to psychological science or to Christian faith—and are doomed to be outdated as psychological science advances further and as biblical scholars offer us new insights into the meaning of scripture. Worthington perceptively put it this way:

> When incongruities arise in the conversation between psychological science and Christian theology—as they inevitably must—we can resolve some simply by checking for discord against *Scripture as we understand it.* But for some

disagreements arising from different findings in theology and psychological science, we must pursue two-sided conversations to try to resolve the discord. We are going to have uncertainties as we try to decide whether to believe a theologian who says one thing or a scientist who says its opposite. (Worthington 2010, 46, emphasis added)

And there's the rub: "Scripture as we understand it." Different ecclesiastical traditions have different understandings of scripture. Advances in biblical scholarship continue to prompt us to revisit some of our traditional interpretations of scripture. Worthington helpfully reminded us of how some strongly held views within certain Christian traditions have changed in recent years when he wrote, "The passages in Paul's letters to the Corinthians on women speaking in church, head coverings, slavery and other issues come to mind. . . . If our understanding of the historical context changes, then our interpretation of Scripture can change" (Worthington 2010, 203). Lesson learned: *stop treating the Bible like a textbook of science, including psychological science.*

We understand the desire by well-meaning Christians to add further credence to what scripture teaches by demonstrating that scripture agrees—is "in concordance with"—what a particular specialized area of contemporary science is currently teaching, whether it be geology, astronomy, evolutionary biology, or psychology. Unfortunately, this approach has repeatedly turned out to be a mistaken path. Advances in biblical scholarship and advances in science will uncover new evidence, and in the light of that evidence the old concordism will unravel. The failure of the concordism efforts gave rise to the integration approach, which we describe in the next two chapters.

# Integration under the Microscope

*Historical Perspective*

---

## THE WIDER CONTEXT:
## INTEGRATION AS A RESPONSE TO PAST FAILURES

ANY MEANINGFUL DISCUSSION of the relationship between psychological science and Christian faith should not be treated as a standalone topic. With such a long and well-documented history of attempts to relate the knowledge given through scripture to that gained through the scientific enterprise, it is necessary to place our own discussions within the context of this wider canvas. There are many lessons to be learned from the past. In earlier chapters, we identified the dangers of contrived conflicts and concocted concordisms as ways of describing the relationship between science and faith. Such attempts, we believe, reflect a failure to grasp the fact that the Bible is not a scientific textbook of astronomy or geology or psychology. These attempts also remind us that scientific explanations of events are not necessarily substitutes for, or competitors with, theological explanations.

It is the failure of past attempts at characterizing the relations between psychology and Christian faith that led to the various formulations of the "integration" of psychology and religion that are considered in this chapter. In particular, the desire for an integrative approach flows from the perception that a deep chasm separates psychology and religion. As mentioned in chapter 2, the hostility toward religion expressed by well-known theorists such as Sigmund Freud and B. F. Skinner gave credibility to the existence of this chasm. But, as also noted, not everyone shared this perception.

## HINTS OF A POSITIVE, CORDIAL RELATIONSHIP

In the century since making its appearance as a scientific discipline, psychology has become one of the most popular undergraduate courses, and psychology books are among the best-selling textbooks. In April 1946 Norman Munn published a book with the title *Psychology: The Fundamentals of Human Adjustment* (Munn 1946). It quickly became a standard textbook for college and university students studying psychology.

Soon after the appearance of Munn's book, another basic textbook of psychology for college and university students was published under the shared authorship of Edwin Boring, Herbert Langfeld, and Harry Weld (Boring, Langfeld, and Weld 1948). This, together with Munn's, were the recommended textbooks for those studying experimental psychology at the University of Cambridge in the early 1950s. Thus, they shaped the way that one of us (MAJ) understood the field of psychology.

What was the attitude of these textbook authors toward religion? Munn's autobiography recounts memories of his early life in South Australia and of "going to Sunday school in the Croydon Congregational church and the joy in singing and knowing that Jesus loved me, because the Bible said so" (Munn 1980, 3). It is clear from reading Munn's autobiography, and from MAJ's personal friendship with him (detailed and documented in this autobiography), that he remained a practicing Christian throughout his life. In addition, one of us (MAJ), during a year spent as a researcher at Harvard, had the privilege of meeting Edwin Boring, professor of psychology at that institution. Knowing of MAJ's interest in the relation of psychology and Christian faith, Boring volunteered during the conversation that he had just written a positive endorsement of a book on science and religion that included a section on psychology.

We could add to this list other leading figures in psychology at that time whose lives and work demonstrated the complete absence of what we earlier described as the conflict motif to characterize the relations between psychology and religion. For example, Gordon Allport, also a professor of psychology at Harvard, published a book titled *The Individual and His Religion: A Psychological Interpretation* (Allport 1950) that exhibited an entirely positive attitude toward religion. One of us (MAJ) was well aware of Allport's personal faith because we both attended the same church in Harvard Square. Finally, we should add that MAJ's mentor at Cambridge, Sir Frederic Bartlett, professor of experimental psychology, made clear his favorable attitude toward religion in his 1950 Riddell Memorial Lectures, published as *Psychology as Experience, Action, Belief* (Bartlett 1950).

Given the very positive attitudes toward religion shown by many leading psychologists on both sides of the Atlantic in the middle of the twentieth century, it seems puzzling that the conflict motif became so widespread within academic psychology, especially in the United States. Why did the conflict motif reemerge? What happened to the earlier amicable relationship? Why did psychology not learn from the mistakes made in applying the conflict model to science-religion discussions in earlier centuries?

These questions are interesting, but the focus of this chapter is not on the conflict motif but rather on the recurring theme of *integration* so prominent in recent discussions. We turn now to a description of the integration approach and our concerns about that approach to the relationship between psychology and Christian faith.

## Defining Integration

### *The Language of "Integration"*

According to Henrika Vande Kemp, a leading historian of the psychology of religion, psychologist Fritz Künkel was probably the first person to use the word "integration" with respect to psychology and theology (Vande Kemp 1996). Künkel, who was an early advocate of Christian counseling, used the term initially in a 1953 letter to a colleague, and then in a 1954 journal article titled "The Integration of Religion and Psychology" (Künkel 1954). Within a few years, the language of "integration" became the dominant way that Christians, especially evangelical Christians in the United States, described the attempts to relate the two domains, and that continues to be the case today.

Although Fritz Künkel is given credit for coining the phrase "integration of psychology and religion," others had expressed a similar idea decades earlier. For example, Olaf Norlie applied the term "Christian psychology" to his efforts to "supplement the science of psychology with the fundamental concepts of the Bible." Norlie claimed that this could be done in a neutral way, "without prejudice to either the scientific or the Christian theological point of view" (Norlie 1924, 25). However noble his attempts to treat psychological science and theology with equal respect, in practice his view of a Christian psychology clearly gave scriptural teachings priority over the results of psychological research:

> This book employs the method and material of philosophy and science wherever these do not conflict with the clear teachings of the Bible, the revealed, inspired, authoritative and infallible Word of God. While the Bible is not a textbook of psychology, it is, on many points, the best and last source: It is an inexhaustible gold mine of psychological information. (Norlie 1924, 25)

Norlie's book foreshadowed an issue that we believe has become a major problem with the integration approach: *the attempts at integration are typically one-sided, in that theological principles take precedence over psychological principles.* Only those aspects of psychology that conform to scripture—as interpreted by the person doing the integration—are allowed into the new, "integrated" model of knowledge about human character and human behavior. This issue comes through strongly in the way that integration is described by three influential figures in this discussion: John Carter, Gary Collins, and Stanton Jones.

## John Carter

In the 1970s, Carter described his view of integration in two articles published in the *Journal of Psychology and Theology* (Carter and Mohline 1976; Carter 1977). He followed these articles with a book coauthored with Bruce Narramore titled *The Integration of Psychology and Theology: An Introduction* (Carter and Narramore 1979). Carter's approach began with the assumption that there is a "unity of truth":

> If we believe that God is the source of all truth, we assume that there is no inherent conflict between the *facts* of psychology and the *data* of Scripture. All conflicts between theology and psychology must, therefore, be conflicts between either the *facts* of Scripture and the *theories* of psychology, the *facts* of psychology and our *(mis)interpretation* of Scripture, or between the *theories* of psychology and our *(mis)interpretations* of Scripture. (Carter and Narramore 1979, 22)

From Carter's perspective, given the unity of truth, there must be a set of shared concepts or principles that are true for both psychology and theology. If and when psychology claims an exclusive right to the truth, integration with theology is required as a corrective step. Thus, Carter said:

> Integration . . . is an attempt to relate Christian thought to the challenge of contemporary culture's attempt to define man according to a secular psychology. This does not imply that psychology is wrong in principle but only that psychology must be integrated with a Christian view of man, his nature, and his destiny. (Carter and Mohline 1976, 4)

After examining twelve books on systematic theology and sixty-five general psychology textbooks, Carter and his colleague Richard Mohline proposed a model that contains what they called the "intrinsic common structure of both psychology and theology" (Carter and Mohline 1976, 4). They claimed that, with the exception of the distinctive methodology used by each discipline, "the basic principles and content of psychology are integratable into their equivalent theological area" (Carter and Mohline 1976, 6).

Thus, Carter's view called for a full and complete incorporation of psychology into theology. All psychological truths can be integrated with the corresponding theological truths, because ultimately, they are the same truths. Interestingly, Carter (1977) contrasted his integration model with three other approaches to the relationship between psychology and religion: *setting religion against psychology* (psychology is superfluous, because all truth is contained within theology); *placing psychology above religion* (psychology has discovered a clear, detailed view of human nature, while theology has a less useful, partial view); and *holding psychology and religion in parallel* (reason and revelation both lead to valid truths, but in different, nonoverlapping spheres).

### Gary Collins

Collins described his view of integration in *The Rebuilding of Psychology: An Integration of Psychology and Christianity* (Collins 1977) and *Psychology and Theology: Prospects for Integration* (Collins 1981). His approach became known more widely as a result of his defense of the "integration view" in the dialogue captured in the book *Psychology and Christianity: Four Views* (Johnson and Jones 2000). Like Carter, Collins held a unified view of truth: God's truth as revealed in scripture is discoverable by psychological research, and truths about human nature that have been established by scientific investigations cannot be contradicted by scripture—any apparent contradictions will eventually be resolved (Collins 2000, 103).

Collins emphatically supported the importance of integration. However, he did not provide a clear definition of "integration" nor a detailed model for how integration should take place. Instead, he stated:

Even after several decades we still cannot define integration in a way that is generally accepted . . . [because] there is no agreement about what we are integrating. To this point I have written mostly about psychology and *Christianity,* but in earlier writings I have discussed the integration of psychology and the *Bible,* and the integration of psychology and *theology* as if these are the same. They are not. . . . Even as psychology is diverse and complex, so are definitions of Christianity, interpretations of the Bible, and Christian theologies. All of this adds to the complexity of trying to define integration. (Collins 2000, 112)

Collins did not see "integration" as merging psychology and theology, or even breaking down the boundaries between them. He wrote:

Do we really want the two fields to become one? For me, the answer is no. . . . Integration implies two separate but unique fields shedding light on our understanding of similar issues. Integration does not imply the disappearance of theology, the elimination of psychology, or the swallowing up of one field by the other. (Collins 1981, 18)

Collins also recognized the dynamic and constantly changing nature of both psychology and theology, as well as the diversity of backgrounds within the group of Christian psychologists. He wrote, "We talk about psychology and theology as if these were two neutral and objective hunks of information that we will somehow, someday, put together. In reality, every effort at integration is a reflection of the integrators" (Collins 2000, 115). In the end, Collins saw integration more as a style of living rather than an intellectual activity: "At its core, integration is a Spirit-led activity and a way of life that starts and ultimately takes place in the mind and soul of the integrator" (Collins 2000, 126).

## Stanton Jones

Jones contributed to the psychology–faith discussion by collecting diverse perspectives on the topics for his edited volume *Psychology and the Christian Faith: An Introductory Reader* (S. Jones 1986). He gained wide attention for an article in the *American Psychologist* that proposed a "constructive relationship" between psychology and religion (S. Jones 1994), and was the representative of the "integration view" in the second edition of *Psychology and Christianity: Five Views* (E. Johnson 2010). In contrast to Gary Collins's claim that "integration is undefinable" (Collins 2000, 112), Jones had a clear working definition of "integration":

> Integration of Christianity and psychology (or any area of "secular thought") is our living out—in this particular area— of the lordship of Christ over all of existence by our giving his special revelation—God's true Word—its appropriate place of authority in determining our fundamental beliefs about and practices toward all of reality and toward our academic subject matter in particular. (S. Jones 2010, 102)

Thus, Jones agreed with the other integrationists in rejecting the idea of psychology and religion as separate, unrelated domains. He stated, "We believe that Jesus Christ is Savior and Lord of all life. We believe that no aspect of life is outside of the scope of his sovereignty." Compared with some of the other advocates of integration, Jones was more open to knowledge derived from psychological research:

> The integrationist also surmises that Scripture does not provide us all that we need in order to understand human beings fully, and that there is a legitimate and strategic role for psychology as a science and as a profession in giving us intellectual and practical tools for understanding and improving the human condition. (S. Jones 2010, 101–2)

This view still gives priority to scriptural principles about human nature. Where there are gaps in the theological explanations, we can fill in the missing information by using the findings from psychological research. But where scripture has spoken, there is no real need for psychology. Even Heath Lambert, an advocate of the "biblical counseling" approach and a frequent critic of Jones, acknowledges that integrationists "have a strong desire to think carefully and biblically about how to *filter out secular visions of life*," and states, "we should admit that we all are trying to *place the Bible in authority over psychology*" (Lambert 2016, 28, emphasis added). Filtering out secular viewpoints can affect not only the interpretation of research results, but also the areas of study. Influenced by philosopher Nicholas Wolterstorff, Jones suggested that a psychologist's Christian commitment will shape his or her "control beliefs," which then may influence which particular areas of psychology the psychologist investigates (Wolterstorff 1984). For example, control beliefs about the importance of helping others may lead one psychologist to study the developmental problems of children or older adults, while beliefs about the importance of understanding God's creation may lead another to study the neural mechanisms that underlie behavior.

The ideas and frameworks that define the *integration approach* were mainly constructed during the 1960s, 1970s, and 1980s. During this period, in addition to Carter, Collins, and Jones, other psychologists supported the goal of integrating psychology and theology, including Paul Tournier, Clyde and Bruce Narramore, James Dobson, Ronald Koteskey, Hendrika Vande Kemp, Larry Crabb, James Beck, and Bruce Demarest. Although there was considerable variation among these individuals, they shared a set of beliefs: (1) integration is necessary because there is a wall between psychology and theology; (2) some psychological theories or research findings conflict with, and threaten, theological principles; and (3) where there are conflicts, theological explanations should take precedence.

## CHALLENGES TO THE INTEGRATION APPROACH

### A Philosopher's Perspective

In his seminal book *Preserving the Person*, philosopher C. Stephen Evans provided a broader view of attempts to link psychology with Christian faith (Evans 1977). Evans began with the idea of human persons as conscious agents with freedom to act according to their values, and responsible for their own choices. Evans proposed that some theories and research findings in psychology, sociology, and neuroscience—such as the ideas of Freud and Skinner mentioned in chapter 2—pose a threat to the idea of personhood, and identified several different approaches that Christians have taken in response to that depersonalizing threat. He labeled those approaches as

- *Reinterpreters* who concede the truth of the scientific view. Some Reinterpreters (the *Capitulators*) believe that personhood is false and must be abandoned, while others (the *Compatibilists*) believe that personhood, properly understood, is compatible with the scientific view.
- *Limiters of Science* who try to maintain personhood by placing boundaries on science. Some Limiters of Science (the *Territorialists*) mark certain topics as "off-limits" for scientific investigations, while others (the *Perspectivalists*) believe that science may legitimately study everything, but is limited to only one dimension or level of reality.
- *Humanizers of Science* who attack the underlying philosophy of science. Some Humanizers of Science (the *Generalists*) reject the scientific method as the only source of truth, while others (the *Particularists*) accept the validity of the scientific method in the natural sciences, but not in the sciences that deal with humans.

Several aspects of Evans's analysis are of particular interest here. Of the Reinterpreters, he wrote:

Time after time, it seems, Christians have opposed new scientific theories because they have unwittingly misunderstood
the essential content of their faith. . . . Basically, the position
of a Reinterpreter is that *the universe is indeed mechanistic
and that there is no reason to think that human behavior is
any exception.* The idea that there are "gaps" in the scientific
chain of causes in which "spiritual" or "mental" activity might
reside is a weak foundation for a Christian understanding of
man. (Evans 1977, 94, emphasis added)

Evans noted that the Compatibilist Reinterpreter position with regard
to scientific determinism (people are free because they could have made
different choices if the prior causes were different than they actually
were) is similar to the view held by some Calvinist theologians about
theological determinism (people are free because they could have made
different choices if God's intention would have been different). On balance, Evans found neither the views of the Capitulators nor the Compatibilists entirely satisfactory. He concluded, "I do not myself regard
the *Reinterpreter* view as a promising option for either the Christian or
the non-Christian humanist" (Evans 1977, 100).

Moving on to the Limiters of Science, Evans wrote:

Limiters of Science believe that the conflict between the image
of the personal and the scientific picture of man is generated
by scientism. *The origin of the problem,* they say, *lies in the
failure to reflect realistically and critically on the nature and
limits of science and the scientific method.* (Evans 1977, 101,
emphasis added)

In this group, he included those he labeled as *Territorialists*—those
who conceive the limits of science in terms of the specific aspects of
reality that science is capable of engaging. Evans wrote, "The classic case

of Territorialism is Descartes's mind-body dualism" (Evans 1977, 102), because dualists hold that science can study the physical body, but not the nonphysical mind or soul. Evans continued:

> In contrast to the dualist, the "Perspectivalist" may stress the unity of man. He may admit that to some extent it is true to say that a man is a body, or that he is a complex of socio-cultural relationships. However, *the perspectivalist stresses that this unitary reality can be seen from different viewpoints, and, when seen from these multiple viewpoints, different aspects appear.* Science is limited not by what sorts of objects can be studied, but by what can be said about the objects it studies. (Evans 1977, 106, emphasis added)

Later in this chapter Evans has a section headed, "Two Perspectivalists: Jeeves and MacKay." Evans referred to the writings of Donald MacKay, and included a quotation (Evans 1977, 110) from a book that one of us (MAJ) wrote in 1976:

> I shall argue that there is no necessary conflict between the assured results of psychological researches into religious behavior and experience, and much of what has been expressed in other ways, using categories variously described as religious, theological, or spiritual about the same behavior and experience. I believe it is making a category mistake to oppose what is asserted in two distinct language domains. (Jeeves 1976, 18)

Moving on finally to the Humanizers of Science, Evans saw this approach occurring frequently in discussions of the human person, especially the Generalist Humanizers who are opposed in principle to "positivist" assumptions in science, arguing that scientific knowledge is inherently flawed. On the other hand, Particularist Humanizers would like to separate so-called hard science from human science, accepting scientific investigations of the natural world but rejecting the legitimacy of the scientific method for studying the human person. In the past few decades, however, much of the human science in psychology has become hard science with the emergence of cognitive science, neuroscience, and evolutionary psychology. After reflecting on these three approaches, Evans concluded that some sort of *perspectivalist model* was, in his view, the most helpful and most faithful to reality.

### Perspectivalism as a Challenge to the Integration Model

As the formative decades for the integration view were drawing to a close, another important book was published in 1986 that serves as a summary and capstone for that era. Pursuing his explorations of how properly to relate scientific psychology and Christian belief, Stanton Jones helpfully gathered together eleven psychologists, theologians, and philosophers to discuss their views on psychology and Christian faith (S. Jones 1986). Following up on the work of Stephen Evans, Jones noted that, in addition to the traditional "integrationists," the field still contained Limiters of Science, including both "territorialists" and "perspectivalists"—such as Donald MacKay and David Myers—who emphasize the need to recognize different "levels of explanation." He quoted one of us (MAJ) as a typical example of a perspectivalist:

> On closer scrutiny, however, sometimes it turns out that supposed conflicts between what psychologists have discovered and what Christians believe have arisen through not pausing long enough to establish precisely what the psychologist is and is not asserting, and what a biblical faith does and does not encourage us to believe. One of the main aims of this book, therefore, is to show how Christian beliefs and the statements of behavioral scientists concerning the same set of events may be related, so that neither is abused and both are given full weight. (Jeeves 1976, 7)

[ 73 ]

The perspectivalist or *levels-of-explanation* viewpoint thus represented one type of challenge to the integrationist approach. Rather than putting theology in a position of authority over psychology, perspectivalists see both domains as contributing valuable, even essential, information about human nature.

Another type of challenge to integration came from Jay Adams, founder of the "biblical counseling" view (Adams 1970), and from Robert Roberts and P. J. Watson, who advocated a "Christian psychology" (Roberts and Watson 2010). Both of these approaches see the gap between secular psychology and theology as too wide for any bridge to bring them together. Therefore, "establishment" psychology must be abandoned in favor of a completely new approach based strictly on the Bible. Integration is not needed, and would be counterproductive, as it would give legitimacy to secular psychology, which has little to offer to Christian faith or to our understanding of human persons and their problems.

### The Multiple Meanings of Integration

Vague and confusing language emerged as another challenge to the integration movement—highlighting a lack of clarity about the definition of "integration" and the specification of what is to be integrated. Steve Bouma-Prediger reviewed the literature on integration as of 1990, listing a mind-boggling array of different meanings for "integration" and different characterizations of the domains to be brought together. He asked:

> What exactly is involved in integration? Does one integrate psychology with faith, or the Bible, or revelation, or theology, or a Christian worldview, or Christian belief, or Christianity, or religion? Does one integrate theology (or faith, Christianity, etc.) with psychology, or science, or therapy, or counseling? Does one integrate theory with practice, or faith with practice, or faith with learning, or faith with vocation, or religious experience with therapy? . . . What exactly does the term "integrate" mean? Does it mean merely to relate, or does it mean, more

specifically, to combine, to harmonize, to unify, or some other possibility? As stated previously, the issue is: what integrates with what, and what is the precise character of that integrative relationship? (Bouma-Prediger 1990, 23)

In an attempt to add clarity to the integration discussion, Bouma-Prediger provided a helpful typology of four basic kinds of integration:

*Interdisciplinary integration* is integration between different disciplines. In this type, foundational issues are examined with the aim of achieving some degree of unity between the two disciplines. *Intradisciplinary integration* is integration of theory and practice within a given discipline or profession. Here an effort is made to direct a specific practice according to a particular theoretical perspective. *Faith-praxis integration* is integration of faith commitment with way of life. In this case, life praxis is guided by religious commitment and world view. Finally, *experiential integration* is integration within the self and between the self and God. Here healing occurs as a result of a religious encounter—for the Christian, an experience of grace. (Bouma-Prediger 1990, 29–30)

Even though the term "integration" applies to all four of these situations, the usage that is most relevant to our discussion is "interdisciplinary integration"—bringing together psychology and theology. In the next chapter, we examine the continuing evolution of the "integration movement" in the twenty-first century. We also describe our objections to the integration approach and lay the foundation for an alternative view of the relationship between psychology and theology.

# CHAPTER 5

# Integration

## Contemporary Views

---

## DIFFERENCES BETWEEN THE US AND UK
## APPROACHES TO INTEGRATION

IN THE UNITED STATES, active members of the Christian Association for Psychological Studies (CAPS) are found mainly in small Christian colleges and universities, less connected to mainstream psychological science. The situation in the United Kingdom was historically different. The formal association of psychologists who are Christians in the United Kingdom started less than twenty years ago as the Network of Christians in Psychology, now called the British Association of Christians in Psychology. On the initiative of Fraser Watts, a psychologist working at the Applied Psychology Unit of the Medical Research Council in Cambridge and a past president of the British Psychological Society, a small initial meeting was held in Oxford. The keynote speaker (MAJ) was a British psychologist who, at the time, was editor in chief of the international journal *Neuropsychologia* and a member of the United Kingdom's main science research funding body, the Science and Engineering Research Council of Great Britain. We mention these professional associations to make clear that both the initiator and the speaker were center stage in mainline developments in psychological science both in teaching and research. This difference in origins and associations, as we shall see later, continues to be reflected in the differing views about integration held by the members of the two bodies.

## SPECIAL JOURNAL ISSUES ON INTEGRATION

The year 1956 witnessed the formation of the Christian Association for Psychological Studies (CAPS) in the United States. To mark the fifty-year anniversary of CAPS in 2006, a special issue of the *Journal of Psychology and Christianity* was devoted to the topic of integration. In 2012, to mark forty years since the establishment of the *Journal of Psychology and Theology*, a special issue of that journal was also devoted to integration. Not surprisingly, there was some overlap in authors who contributed to these two special issues.

We shall review the approach taken by these two journals, both in these special issues and in other representative articles published in these journals over recent decades. Although our review does not claim to be comprehensive, we believe it gives a faithful representation of the recurring views of integration expressed in these journals. The first thing we note is that, despite the 140 years of psychology as a scientific discipline, there are very few references in these journals to the work of historians on the relationship between science and faith, such as Ian Barbour (2000) or John Brooke (1991). This is an important omission, because there are lessons to be learned from other disciplines which, at the time, struggled to establish a clear claim to being a science. Attention to this extensive literature would have given historical depth and helped to avoid some of the errors in the use of the concept of *integration* (de Felipe and Jeeves 2017).

However, in fairness, it could be argued that there were reasons for the narrow focus of most contributions to these journals. First and foremost, most of the members of CAPS—and most of the subscribers to these two journals—are Christian psychologists specializing in pastoral psychology, counseling, and psychotherapy. Moreover, most of the authors of articles in these journals hold academic positions at Christian colleges and universities, Bible schools, or theological seminaries. Representation from psychologists working in mainline American universities is very meager indeed. This, we believe, is a worrying feature if one wishes one's writings to help Christian students studying at secular institutions and struggling with some of the apparent challenges to their faith from the content of their psychology courses.

## The Journal of Psychology and Christianity

As previously mentioned, the *Journal of Psychology and Christianity* published a special issue in 2006 on the theme of integration. The "table was set" in an article by Yangarber-Hicks et al. (2006), which described several features of the current scene that are especially relevant to this book (highlighted with our emphasis added throughout the quotations below). In the opening section of that article, Natalia Yangarber-Hicks wrote:

- ▶ "The pursuit of integration between Christian faith and psychology in the last 50 years has produced much fruit *despite the lingering lack of consensus regarding definitions of the integration enterprise*, criteria for determining truth, and future directions."
- ▶ That "those committed to integration have much work left to do."
- ▶ "Even a cursory survey of literature in the field suggests *that important voices and perspectives on Christian psychology have either not made their way to or have been infrequent guests at the integration table.*" (Yangarber-Hicks et al. 2006, 338)

Continuing her analysis of the current scene, she helpfully reminded us:

- ▶ "An examination of existing literature on integration suggested that clinical and counseling branches of psychology have received a disproportionate degree of attention."
- ▶ That this first point was previously underlined by Hendrika Vande Kemp (Vandekemp, 1988, 202), who noted that "one of the biggest weaknesses of contemporary Protestant Christian psychology is its nearly exclusive emphasis on psychotherapy and the concomitant disinterest in general psychology."

▸ That we need to take account of "the challenge posed to all subdisciplines of psychology by the recent explosion of neuroscientific research [which] will require significant reflections and responses from Christians working in these areas"—a point repeatedly made in this book. (Yangarber-Hicks et al. 2006, 339)

The second section was contributed by Charles Behensky, who wrote from the viewpoint of a specialist in cognitive neuroscience. He emphasized the importance of paying equal attention to "bottom-up" processes (neural activity that gradually gives rise to conscious experience) and "top-down" effects (conscious expectations and goals that guide the activity of lower-level neural networks). It is noteworthy that at the end of his contribution he repeatedly referred to *enrichments* and *insights*. For example, at the beginning of his final paragraph he wrote:

Ultimately, cognitive neuroscience provides a *richer understanding* of our human experience. It is necessary, but not sufficient as an explanation. We *gain an insight* into the development of problematic behaviors and can view treatment as a way of altering our neural connections, without necessarily specifying the best approach. (Yangarber-Hicks et al. 2006, 341)

It is also interesting that in his final sentence he used words that suggest what we have elsewhere described as a *perspectivalist* or *levels-of-explanation* approach when he wrote, "There is value to be found in *several perspectives*, and our task is to identify the points of commonality and determine how the findings of one field can *complement and inform* those of another" (Yangarber-Hicks et al. 2006, 341).

Another important contribution to this discussion was the section by Nicholas Gibson on UK perspectives on the integration of psychology and Christianity. Gibson noted that:

*The dominant approach to the integration of psychology and theology in the U.K. has much in common with dialogical work on the interface of science and faith.* Rather than trying to subsume one discipline within another or merge them to create a "Christian psychology," the approach has been to set up a dialogue between theology and psychology at multiple points of interface. (Yangarber-Hicks et al. 2006, 346)

He went on, "Besides the obvious interface between Christian pastoral care and mental health, each discipline can offer *complementary perspectives* to the other regarding non-clinical concerns" (Yangarber-Hicks et al. 2006, 346). As one example, he referred to Fraser Watts' book *Theology and Psychology* (Watts 2002):

Here Watts considers how theological reflection on topics such as evolutionary psychology, consciousness, artificial intelligence, and the self *can enrich* psychological understandings of human nature and how *psychological perspectives* on doctrinal topics such as divine action, salvation history, and eschatology *can enrich* systematic theology. (Yangarber-Hicks et al. 2006, 346)

A final very important point was emphasized by Gibson:

This distinctive of British integration work reminds us of *the importance of integrating the full scope of basic psychological theory and research* with theology in all its breadth. While many people may first approach integrative issues from an applied perspective, *there is no good reason why integration as a field should maintain its current lopsided focus on clinical and pastoral issues.* (Yangarber-Hicks et al. 2006, 346)

Sadly, our analysis of recent writings on integration in the United States suggests that Gibson's plea remains "a voice crying in the wilderness."

Another relevant paper in this special issue provided a wide-ranging scholarly review of views on integration in the intellectual marketplace at that time. This conversation was moderated by Elizabeth Lewis Hall (Lewis Hall et al. 2006). It is noteworthy that several of the contributors expressed their unease with the use of the word "integration," given its multiple and diverse understandings by different people in different contexts. For example, Mary Stewart Van Leeuwen opened the conversation with a blunt statement: "My biggest challenge has been trying to get rid of the word integration. I don't like it because I prefer to talk about the unity of faith and learning" (Lewis Hall et al. 2006, 332). She continued:

> I don't like the word integration because everyone reasons within the bounds of religion, to borrow Nicholas Wolterstorff's (1984) book title. And the very fact that we use the term integration is, as far as I can see, an indication of how much we are affected by the enlightenment mindset that the world consists of brute facts . . . and values only appear on the scene when a human being appears to have them. (Lewis Hall et al. 2006, 332)

Another contributor to the conversation, Bruce Narramore, expressed similar unease with the term "integration" when he said, "I don't feel comfortable, I actually never have, with the word integration, for some of the reasons that Mary enunciates" (Lewis Hall et al. 2006, 333).

In her short postscript to the conversation, Elizabeth Lewis Hall wrote, "The challenge to integration is to remain in conversation, not to endlessly rehash old issues, but to reformulate them in increasingly sophisticated ways that reflect developments in our field and our understanding of faith" (Lewis Hall et al. 2006, 336). Regrettably, Lewis Hall's high expectations have yet to be realized. In fact, we find, especially among psychologists interested in counseling, pastoral psychology, and clinical psychology, a constant rehashing of the same old issues. By focusing only on these areas, those who write about them have failed to follow Lewis Hall's exhortation to reformulate these issues in increasingly sophisticated ways—ways that are faithful both to psychological science and to our understanding of scripture.

As a result, large areas of contemporary psychology as studied by students at colleges and universities are seldom discussed by those interested in the integration issues. Moreover, seldom do we find references to the relevance of advances in biblical scholarship in the attempts to relate what we learn of creation—including about ourselves—from psychological science and what is revealed to us in scripture. James Beck lamented both aspects of this situation in an article looking ahead to the next fifty years of integration attempts (Beck 2006). He predicted that, in the future:

> Christian psychology will display an increased need for sophisticated empirical research to undergird its efforts to deliver quality services. . . . Without this research base, the Christian integration movement runs the risk of being crowded out of the academic world and of being relegated to history's dustbin of ineffective treatment regimens. . . . Integrators must likewise upgrade the sophistication of the biblical and theological materials they utilize in their work. A weakness of the integration movement to date is that we have not extensively involved biblical and theological scholars in our enterprise. (Beck 2006, 328)

### The Journal of Psychology and Theology

As previously mentioned, in 2012 the *Journal of Psychology and Theology* published a special issue devoted to the topic of integration. The editorial introduction to this issue, written by Everett Worthington and Todd Hall, emphasized the extent to which psychology has evolved, forcing an evolution in approaches to relating psychology and theology. They wrote, "The world has changed. Integration also has changed. In fact, integration is much more diverse now than the virtual mono-approach of 40 years ago" (Worthington and Hall 2012, 3). These authors also noted the diversity of views present in this special issue.

The issue included contributions from two Baylor University philosophers, Stephen Evans and Robert Roberts. These papers remind us that philosophers have been reflecting upon human nature for centuries, and that philosophers contributed to the development of the discipline of psychology. For example, although Roberts focused his discussion on the ideas of psychologist Paul Vitz, he nevertheless pointed out the important contribution made by Aristotle in his writings on the nature of human flourishing, human upbringing, human action, and human friendship—topics still of major interest to twenty-first-century psychologists (Roberts 2012). In a similar vein, Evans (2012) encouraged us to draw upon the contributions of thinkers such as Augustine, Aquinas, and Kierkegaard. However, we should keep in mind how difficult it must be for a philosopher, peering in from the outside, to properly understand what psychologists are doing. Psychology is not a spectator sport. It is a science that requires deep immersion in the details; within even a small research topic there is a vast literature and there are many specialized skills to be learned and applied. This may help to explain an occasional lack of understanding between psychological scientists and the philosophers commenting on the research findings.

Roberts (2012) urged us to face the challenges to integrate science and religion across time as well as across disciplinary content. This was why, in earlier chapters, we reviewed the lessons to be learned from the past about the dangers of generating *pseudoconflicts* and *false concordisms* in the relations between science and faith. However, it is not clear that Roberts has learned from the errors of past attempts at *integration*—for example, attempts to derive principles of astronomy from scripture simply because biblical authors refer frequently to the stars, the moon, and the earth. There is no "Christian astronomy," and, contrary to Roberts's position, we would argue that neither is there a "Christian psychology."

In a different context, Everett Worthington, one of the coeditors of this special issue, urged caution about the use of the word "integration" when he wrote:

> Indeed, there are times when I think integration is appropriate. But I must admit that there are many areas that theologians address that psychological science cannot. I cannot study eschatology by conducting psychological experiments. I cannot address history scientifically but have to rely on people's memory of history. (Worthington 2010, 104)

Evans voiced a similar thought:

> With all due respect to the *Journal of Psychology and Theology,* I think much harm is done if we think of integration primarily as an attempt to integrate the two disciplines of psychology and theology. . . . Obviously, this does not allow Christianity to play much of a substantive role within psychology. . . . In reality, Christian faith has much to offer the psychologist as a psychologist. (Evans 2012, 32–33)

It is significant that Fraser Watts, psychologist and theologian, determinedly avoided talk of *integration* in his contribution to this special issue. Instead he wrote, "Though the claims of the two disciplines are often seen as alternatives, it is argued that they are better seen as *complementary perspectives*" (Watts 2012, 45). Not only did he see them as complementary, but he also saw the possibility of *mutual enrichment* between the two disciplines. This, he believed, can be brought about by greater dialogue. For example, he wrote, "There is also too much generalization about human nature, and neglect of differences between people. If theological anthropology was more interdisciplinary, and conducted in dialogue with an empirical discipline like psychology, the result would be more satisfactory" (Watts 2012, 47). Dialogue, complementarity, and enrichment are the key notions for Watts, and with this we would agree.

Australian psychologists Maureen Miner and Martin Dowson, after briefly reviewing the possibilities of how to relate psychology and Christian faith, made it clear that their choice was complementarity. They wrote:

> A third approach, complementarity, moves away from the integration of disciplines to the integration of explanatory accounts of particular phenomena. . . . This present paper will use the method of complementarity to generate this form of integration of psychology and theology. . . . The limitations of psychology and theology alone in explaining spiritual experiences suggest that both are needed in order to provide a more complete account of experiences of relationship with God. . . . Complementarity allows for each discipline to contribute to an inclusive *explanatory* synthesis without violating either discipline. (Miner and Dowson 2012, 56, 58)

We support the complementarity approach referred to here, in that the statements made from the two complementary perspectives of psychology and theology may together give a fuller account of what is being studied and therefore may be *mutually enriching*.

Most of the remaining contributions to this special issue on integration are autobiographical. It is, of course, interesting and informative to hear of the pilgrimage that individual Christians have taken in their psychology careers. What is perhaps regrettable is that this special issue contained so few critical, analytical contributions that take apart and examine in detail the whole concept of integration.

It is noteworthy once again that the majority of contributors have a psychological specialization in clinical psychology, family psychology, and psychotherapy. Why were there no authors who have made significant contributions in the literature of, for example, evolutionary psychology, neuropsychology, the psychology of perception, or evidence-based psychopathology? In our opinion, restricting the discussion of integration to only a few areas of psychology is one of the major obstacles to progress in the relationship between psychology and religion. Surprisingly, none of the contributions to this special issue addressed this concern directly.

One contributor, Alan Tjeltveit, did briefly touch upon this question when he reminded us of the sustained scholarly work of Paul Meehl, a former president of the American Psychological Association, and his colleagues in their comprehensive 1958 volume *What Then Is Man?* Tjeltveit wrote:

> Meehl et al. (1958) eschewed easy answers to integrative questions. "There is a need for hard-headed, rigorous analysis of the conceptual relations involved," they asserted. "Psychology and psychiatry are complex (and primitively developed) sciences; theology is complex and, by its very nature, full of obscurity and mystery. It can be assumed that the conceptual relations that arise when those two are brought together will have a high order of complexity." (Tjeltveit 2012, 302)

An article by Keri Barnett and her colleagues in the very next issue of this journal called attention to the fact that the contributors to the special integration issue are relatively isolated from the main body of psychologists, and noted that this applies generally to contributors to the *Journal of Psychology and Theology*, who tend to come from the same "club" with similar institutional affiliations. Barnett and her colleagues wrote:

> One puzzling finding was the lack of change in the institutional affiliations of most authors for the most cited integration articles over the last quarter of a century. On one hand,

one would assume that integrationists would want psychology in general to have an increasing interest in integration, as evidenced by greater authorship from their secular counterparts, as well as having more secular institutions employing explicitly Christian integrationists. On the other hand, it is also reasonable to expect that integrationists affiliated with Christian institutions should be stepping up to the plate and contributing at least as much as their secular counterparts. Although there are more secular institutions in comparison to Christian institutions, the faith integration emphasis inherent to most Christian institutions makes it especially important that religiously affiliated faculty members are contributing to the integration literature. However, neither of these is clearly happening, which begs the question: How much does the discipline want to be internally versus externally driven and focused? The direction of future growth in the integration field will partially depend on how this question is answered. (Barnett et al. 2012, 98)

### Integration's "Walled Garden" Problem

Sadly, the pattern of articles over the years since that special issue does not appear to reflect an expanded, more inclusive perspective—an attempt to break out of the "walled garden" of counseling and psychotherapy. This is worrying and brings to mind David Myers's response to two chapters in the second edition of *Psychology and Christianity: Five Views* (E. Johnson 2010): a chapter by Robert Roberts and P. J. Watson, who advocate a "Christian psychology" (Roberts and Watson 2010), and David Powlison's chapter outlining a "biblical counseling" view (Powlison 2010). Both of these chapters advocate an outright rejection of mainstream (secular) psychology. In response to Powlison, Myers wrote, "This leaves me little to say, other than that when he and I use the word *psychology*, we are not talking about the same thing" (Myers 2010d,

275). In response to the "Christian psychology" view, Myers makes clear that he is deeply unhappy with such an approach. He writes:

> Rather than replacing psychological science with the sages of the ages, why not respect both as valuable and limited? . . . I find myself intrigued by psychological science's discoveries, which the sages of the ages seem not to have anticipated. (Myers 2010b, 180)

He then, by way of illustrating psychological science's usefulness, listed some findings from dedicated empirical research in psychology over the last half-century. He wrote, "If establishment psychology is indeed where significant discoveries and new understandings are emerging, do we really want to run off into the corner to create our own Christian psychology? By doing so, do we not risk irrelevance?" (Myers 2010b, 181). He quotes C. S. Lewis, who once declared, "We do not need more Christian books; we need more books by Christians about everything with Christian values built-in" (Myers 2010b, 181).

Myers continues, "Or as Stanton Jones says in his chapter, 'Christians should be in the thick of psychology, contributing their ideas, submitting their hypotheses and theories to the test'" (Myers 2010b, 181). Myers's response raises an issue that crops up repeatedly in the contributions to both of the special issues on integration: How can the integrationists at Christian colleges, universities, and seminaries be encouraged to engage *all of psychology*—not just counseling, clinical, and pastoral psychology—including *up-to-date findings of psychological science*?

On this point, Everett Worthington, with his skills both as both a counselor and a psychological scientist, comments on the problem of outdated science. He writes:

> Another reason for the weaknesses in the integration by clinician-theologians is that clinicians, looking for ways to

harmonize their psychological theory and theological presuppositions, sometimes adopt theories of therapy that are not current. Presuppositions may trump data on the effectiveness of the theory. (Worthington 2010, 127)

The past decades of integration work demonstrate the *temptation to "harmonize" psychological theory and theological presuppositions.* Apparently, given the prestige and authority of science in the twenty-first century, some Christians believe that, if they can show their theology "harmonizes with" or "fits with" science, it makes their theology more believable. But, even if such a concordism seems persuasive today, those same theological explanations will be discounted later when the science that it "fits" today has been superseded.

## Diverse Views of Integration

Our brief review has revealed considerable diversity in the use of the term "integration" when discussing the relationship between psychological science and Christian faith. We have observed this diversity in two journals specifically devoted to the psychology-Christianity relationship, as well as in books written or edited by leading proponents of integration. We identified views ranging from advocating total intermixing of psychology and theology to an explicit rejection of the whole idea of integration, seeing it as a misguided endeavor.

We summarize this diversity of views in the following categories:

- ► *Total integration*—a complete intermixing of psychology and Christianity, producing an amalgam that can be labeled "Christian psychology." Eric Johnson (2011) is a proponent of this view.
- ► *Selective integration*—to be sought where the subject matter makes it appropriate, especially on topics for which scripture provides very little information. Stanton Jones (2010) is a good example here.

Nevertheless, on close examination this approach often leads to some form of an amalgam.

- *Limited integration*—to be sought where the structure of the theological principle and the psychological theory are similar, but only justified following "a hardheaded rigorous analysis of the conceptual relations involved" (Meehl et al. 1958, 302; Tjeltveit 2012). Stephen Evans (2012) seems to be a champion of this view, although it is evident that he himself is all too aware of the dangers at hand when integration gets out of control.

- *No integration*. We should avoid attempts at integration altogether because integration is totally inappropriate. This position is advocated by those who believe that secular psychology has nothing to offer, such as followers of the "biblical counseling" approach of Jay Adams (Adams 1970). Interestingly, the term "integration" is also rejected by some individuals who accept the relevance of psychological research. As noted above, both Mary Stuart van Leeuwen and Bruce Narramore have stated their reservations about the word "integration."

Given this range of views, the question must be raised, "Is there a better way of thinking about how to relate the knowledge gained from psychological science with that which comes from our Christian faith?" It seems clear that many of the confusions and misunderstandings about integration could be avoided if all discussions on this topic could begin with some common starting points, agreed upon in advance. Two such points come to mind:

- The authors or discussants should set out clearly their perspective of what is and is not included under the rubric of psychology. One way of doing this would be to take the chapter headings in a standard textbook of psychology and indicate which of the subareas of contemporary psychology will be considered in their discussions of the integration of psychology and Christian faith.

▶ The authors or discussants should then state explicitly the reasons for choosing to talk about certain areas of contemporary psychology *and to exclude others*. In this way, the reader will be aware of the limited applicability of any conclusions drawn.

## WHERE NEXT?

As we have mentioned before, any attempts to integrate psychology and Christian belief must deal with psychological science (including neuroscience) *as it actually exists today*. The temptation has always been to assume that the state of psychological research and theory at any particular time is the final word. The way that such an assumption can lead one astray is vividly illustrated by a quotation from Gary Collins in a 1972 review of a book about Abraham Maslow. He stated, "We need a psychology with new dimensions—a psychology which is developed and tested against the truths that are proclaimed in the divine Word of God" (Collins 1972). One wonders how Collins's views would sound today in the ears of Virginia Holeman, associate dean of the School of Theology and professor of counseling at Asbury Theological Seminary. In a chapter titled "The Neuroscience of Christian Counseling," she discussed in detail what contemporary neuroscience can contribute to Christian counseling. Holeman concluded that

> by taking the neuroscientific tightening of the mind/brain connection to heart, one can emphasize the enhancement of clients' relational capacities within a therapeutic setting, superintended and enabled by the agency of the triune God, and embodied in counselors' and clients' relationships with God and through the therapeutic relationship itself. (Holeman 2004, 157)

Given the importance that Holeman establishes of the relevance of developments in neuroscience to counseling, the question must be asked, with Gary Collins's review in mind, *Which bits of psychology and neuroscience are to be tested against "the truths that are proclaimed in the divine Word of God?"* The scriptures, we would argue, were never intended to teach psychology or neuroscience. Therefore, it is meaningless to claim that you can test the truth of these disciplines against scripture, and perhaps even worse, it is demeaning of the purpose of scripture itself.

Psychology also draws heavily on mathematics. For example, some of the most powerful explanatory psychological models of visual perception are expressed in terms of mathematical equations. It is difficult to see how we would "integrate" Christian beliefs with mathematical equations. It's not likely that anyone has ever suggested that Christians who are studying psychophysics should add another constant to Weber's Law to make it more compatible with scripture. The take-home message is clear: let scripture speak what it is given to reveal. Don't pretend that hidden within scripture, if only we could discern them, are modern scientific theories of human nature.

This reminds us that one of the commitments a Christian takes on is a commitment to truth. To us, this means telling the story about the world—including ourselves and our brains—as it really is. In discussions of integrating psychology and Christianity, more attention needs to be paid to evaluating the evidence for each position, in light of the ever-changing findings of psychological science and the evolving interpretations of scripture.

## Taking Stock: Lessons Learned

When thinking about how best to characterize a constructive relationship between psychology and Christian faith, we admit that the widely used concept of integration has definite appeal. In fact, when carefully defined, it seems to have served us well. Nevertheless, as Stephen Evans (Evans 1977) and Everett Worthington (Worthington 2010) have reminded us, the concept of integration is not without its problems.

In our review of both journals' special issues on integration, as well as a selected sample of papers in the *Journal of Psychology and Theology* and the *Journal of Psychology and Christianity* over several decades, we found that the range of views of the concept of integration represented in, and typical of, contributions to these journals is so great that no brief treatment could hope to capture all its richness and diversity. However, our examination of recent relevant articles reveals some common themes.

First, there is a strong and recurring theme of contributions particularly focused on counseling and psychotherapy. Eric Johnson is a high-profile example of this genre. For example, in one of his papers on integration he writes, "In this essay we will consider three of the most important kinds of integration practiced among Christians in psychology, psychotherapy and counseling" (E. Johnson 2011, 339). This represents a "restricted view" of the relationship between psychology and religious faith; the paper makes almost no reference to 90 percent of the psychology taught in a typical undergraduate college or university course. Hence a Christian student taking psychology courses will get little help in connecting psychological science and Christian beliefs from articles like this one.

A second theme, also involving a "restricted view," is the danger of unexamined assumptions about scripture, psychology, and the nature of integration. For example, this issue was noted in a paper by Brad Strawn, Ronald Wright, and Paul Jones. These authors make the important point that we all come to the integration table deeply steeped in our own ecclesiastical traditions with their often-undeclared presuppositions (Strawn, Wright, and Jones 2014).

A third theme and recurring problem is a general lack of historical perspective. For example, it is unfortunate that even a paper specifically focused on integration (Beck 2006) makes no reference at all to the very important and enduring contribution by Evans (1977), which provided a detailed review and analysis of how the concept of integration was being used by various authors in the formative era of the integration movement. An exception to this trend, noted earlier, was an article by

Lewis Hall et al. (2006), which contained a scholarly analysis of a wide range of views on integration.

## Time to Retire the Term "Integration"?

We conclude that, while the concept of integration has at times served us well, perhaps the time has come to drop the word "integration" altogether in favor of language that more clearly and precisely describes what we are trying to accomplish. It will not surprise the reader to learn that we lean toward the approach that Evans (1977) has called *perspectivalism* and that others label the *levels-of-explanation view*. Although we acknowledge the need to develop a crisper, catchier term for this approach, we believe that the approach itself is sound and offers the best foundation for moving forward in a productive way. We discovered (with some satisfaction) that support for the *perspectivalist* position shows up frequently in the special issues of the journals mentioned above, and in books written from an *integrationist* point of view. Even an author like Gary Collins, a well-known advocate of the integration approach, at times seems to echo the perspectivalist view:

> What, then, is the purpose of integration? First, theology and psychology together can ask questions of each other and share perspectives that can stimulate research and lead to a greater, clearer discovery and comprehension of God's truth (especially about people). Second, the task of integration keeps the channels of communications open between theologians and psychologists so that respective conclusions are used to help all of us more fully understand human beings. (Collins 1981, 18)

Perhaps Christian psychologists could adopt a less defensive posture and a more positive attitude by focusing less on attempts to build a complete, integrated system of "psychology-theology" or "Christian psychology," and more on looking for points of contact between the two disciplines that can provide mutual insights and enrichments from these different perspectives.

## LOOKING AHEAD TO THE SECOND HALF OF THIS BOOK

In the second half of this book, we provide examples of the *perspectivalist* approach as it is actually practiced by psychological scientists who are committed Christians. Most of those examples involve *psychologists* applying research findings to illuminate theological concepts. However, we believe that *theologians* can also gain something from applying psychological science to their work. In this regard, Joanna Collicutt, psychologist and Anglican priest, has recently spelled out in detail, with full documentation, what "academic psychology"—her term for mainstream psychological science—could contribute to the study of the Bible. She wrote:

> Academic psychology concerns itself with the scientific study of behavior and mental processes of human beings in interaction with the environment, or to put it less technically, the study of people. It aims to give a rigorous account of how people come to feel, think, and act as they do. In this it clearly interacts, and may at points overlap, with several other human sciences such as anthropology and sociology. While these other human sciences have been readily utilized by biblical scholars, the literature on biblical studies indicates a marked avoidance of psychology, and the occasional expression of an outright aversion to it. (Collicutt 2012, 1–2)

She found this surprising, given that psychological principles seem directly relevant to the behaviors described in the Bible:

> This [avoidance of psychology] is odd, for at first glance psychology appears to be an academic discipline whose applica-

tion to the study of sacred texts, including the Bible is both appropriate and potentially highly productive. . . . Secondly, the content of most sacred texts, especially sacred narratives, can be said to be dominated by psychological themes. These texts contain accounts of human behavior and divine behavior described in anthropomorphic terms that proceed according to certain rules. Motives are attributed to the protagonist, and inner thought processes are sometimes made explicit. . . . Thirdly, sacred texts and human psychological processes seem, at least on the face of it, to be intimately related to each other. Sacred texts are written (if not authored) by human beings and read by human beings. (Collicutt 2012, 2–3)

Collicutt pointed out that the limited use of psychology in the biblical studies literature comes almost entirely from the "depth psychologies" of Freud, Jung, and Erikson. From our perspective, there is a remarkable parallel between the way biblical scholars have tended to focus on depth psychologies and the fact that so many of the leaders of the "integration movement" are counseling psychologists at Christian colleges, some of whom still depend upon these depth psychologies. Collicutt noted, "Thus the psychologies that have been of primary interest to biblical scholars are generally not the psychologies studied by present-day psychologists and vice versa" (Collicutt 2012, 6). This echoes what we have said several times: the primary interests of psychologists contributing to the "integrationist" journals and books are generally focused in areas that are rarely studied by present-day psychological science.

Collicutt (2012, 7) argued that biblical scholars need to tap into the full range of mainstream psychologies that include

- ▸ *Cognitive psychology*: the study of human thought processes [including memory], using models based on information processing theory.

- *Social psychology*: the study of how human beings relate to each other in groups of varying sizes, often using systems theory.
- *Developmental psychology*: the study of human change and continuity over the lifespan.
- *Personality psychology*: the study of differences between individuals, which may include differences in aptitude.
- *Evolutionary psychology*: the study of evolutionary origins and survival value of human behaviors.
- *Biological psychology*: the study of the relationship between brain and behavior (neuropsychology), and between bodily states and behavior.

Collicutt concluded:

> It has been argued that psychology is fundamentally about people and how they actually behave—how they think and remember, make and conserve meaning, attempt to communicate, process and express emotion, and exist in relation to others. Any approach to the Bible that does not pay due attention to the human factors, doing them the respect of applying rigorous methods to their study, is likely to be impoverished. (Collicutt 2012, 48)

We go on to provide examples of how approaches like Collicutt's, which engage psychology as it is actually practiced by psychological scientists today, have the potential to offer a remarkable range of new insights into human nature, enriching our understanding of what it means to live as a Christian in the community of the church.

# CHAPTER 6

# Insights from Neuropsychology

## *An Overview*

W E BELIEVE THAT much of the confusion about properly relating psychological science and Christian faith stems from a failure to recognize the importance of the recent advances in neuroscience for both research and practice in psychology. This accumulated neuroscience evidence is relevant to the work not only of neuropsychologists and cognitive scientists but also of counselors, clinical psychologists, child psychologists, and psychotherapists. As we have seen, this last group of specializations is very well represented in the membership of the Christian Association for Psychological Studies (CAPS) and figures prominently in many discussions of psychology and faith relations.

The interest in the topic of this chapter was underlined recently by the Wellcome Trust, the largest supporter of medical research in Britain, which organizes regular displays at its London headquarters on issues deemed to be of popular and widespread interest. In 2016, they presented an exhibition titled *Mind and Body*, introduced with these paragraphs:

> Why do we talk about mind and body rather than brain and body? Perhaps because we see the brain as a physical entity, controlling the nervous system and bodily functions while the mind holds memory, emotion and personality. In the past, society might have considered the essence of what makes an individual the "soul," and the seat of the soul hasn't always

been considered to be the brain. The Babylonians thought that human emotion and spirit resided in the liver; the Egyptians, that the soul (or *ka*) was located in the heart. The Mesopotamians hedged their bets with the theory that intellect lay in the heart, emotions in the liver and cunning in the stomach.

Are the mind and the body separate? Western medicine seems to have come full circle, beginning and ending with an approach that recognises the interdependency of mind and body, bracketing several hundred years when the two were approached as distinct entities. "It's all in the head" seems to have faded from popular parlance over recent decades as we come to understand that the vast majority of afflictions have both mental and physical aspects. (Wellcome Collection 2016)

While that quotation reflects a view widely shared among professionals treating psychological disorders, we suspect that the extent of this interdependency of mind and body—in our conscious thoughts, our emotional reactions, our ethical and moral decisions, and our religious behavior—is not so widely understood. In this chapter, we present some evidence from neuropsychology and neurology that demonstrates our psychobiological unity as persons.

## An Explosion of Neuroscience Research

Given the volume and intensity of research in neuroscience over the past half-century, it is perfectly understandable that those not specializing in neuroscience simply cannot stay up-to-date with what has been happening. It is therefore necessary to highlight some important developments and discoveries in research at the interface of neuroscience and psychology in recent decades.

In the seventy years since David Hubel and Torsten Wiesel received the Nobel Prize for their work in neuroscience, the field has experienced a quite breathtaking expansion of knowledge about our remarkable human nature. The speed of expansion of the field is well illustrated by the fact that at the inaugural meeting of the Society for Neuroscience in 1969 there were a hundred participants, while today at similar meetings there are more than thirty thousand. Funding of research in this area by governments and charities has expanded at a rate paralleling the increase in the number of researchers. As another indication of the importance of this area of science, the US Congress designated the 1990s as "The Decade of the Brain." Others have called the first decade of this century "The Decade of the Mind." More recently, we have the "Brain Initiative" in the United States announced by President Barack Obama in 2013 and the "Human Brain Project" approved by the European Union in the same year.

These rapid developments and their wider implications underline the wisdom of David Hubel's 1979 statement that "fundamental changes in our view of the human brain cannot but have profound effects on our view of ourselves and the world" (Hubel 1979, 52). As an example of the prophetic nature of this comment, Francis Crick, another Nobel laureate and codiscoverer of the structure of DNA, gained wide attention when he published a book speculating about the implications of developments in neuroscience for traditional and widely accepted religious beliefs about the human person—such as whether humans possess an immaterial and immortal soul (Crick 1994). The publicity surrounding this book served to revive the science and religion conflict motif in a new generation.

For the benefit of the nonspecialist, we review some major features of the contemporary neuroscience landscape, noting especially those findings that seem to have wider implications for some Christian beliefs, such as

- ► Whether evidence of the intimate interdependence of brain and mind reopens earlier debates about dualism and monism.
- ► Whether neuroscience findings have invalidated some central aspects of theology, as Crick (1994) claimed: "The idea that man has a disembodied soul is as unnecessary as the old idea that there

was a Life Force. This is in head-on contradiction to the religious beliefs of billions of human beings alive today."

▶ Whether there are limits to our capacity to act freely, whether brain activity can be monitored to discover an individual's intentions before an action occurs, and whether these findings reduce a person's culpability for immoral actions.

▶ Whether findings about the progression of Alzheimer's disease as it systematically destroys aspects of life (including spiritual life) may provide clues about the way that other forms of brain damage can undermine moral reasoning and moral behavior.

## Some Historical Perspective

The landscape of neuroscience is so vast that the challenge is to highlight the salient features without distorting the overall picture. In his paper on the legacy of Franz Joseph Gall, neuropsychologist Stuart Zola-Morgan divided the history of ideas about localization of brain function into three eras (Zola-Morgan 1995). The first era, from antiquity to the second century, was focused on debates about where the soul was located. The soul here referred to the essence of a person's being and was the source of all mental life. Early Greeks thought the soul was contained in several parts of the body, including the heart and the liver, but that part of the soul that controlled mental functions was generally thought to be located in the head. It was Galen, an anatomist of Greek origin, who did the most to advance a scientific understanding of the soul's mental functions. He performed experiments on animals as well as treating gladiators who had suffered head injuries, and made a strong case for connecting mental functions specifically with the brain.

Zola-Morgan's second era runs from the second century to the eighteenth century. During these years, the debate centered on whether cognitive functions were localized in the "gaps in the brain" (the fluid-filled ventricular system) or in the brain tissue itself. The medieval church exerted a strong influence on this debate by taking the position that ethereal spirits and ideas flowed through the empty spaces of the brain. By the fifteenth and sixteenth centuries, some researchers, including Vesalius, began to question this idea. In the following century, Thomas Willis, who coined the term "neurology," paved the way for the study of brain tissue. In his *The Anatomy of the Brain* (Willis 1664), he argued for a distinction between the immortal "rational soul" that was unique to humans and the "corporeal soul" shared by humans and other animals. This distinction allowed Willis to avoid clashing with contemporary ecclesiastical authority. Having publicly acknowledged the distinctly human "immortal soul," he was free to adopt a psychophysiological approach to the "animal soul" embodied in the brain tissue. As Paul Cranefield wrote, "The soul of brutes, in the hands of Willis, really seems to be simply a handy name for the assemblage of anatomical and physiological mechanisms which underlie psychological processes" (Cranefield 1961, 306).

The principal concern of this chapter lies with Zola-Morgan's third era, from the nineteenth century to the present. During this period the debate has centered on how mental activities are organized in the brain. One early view claimed that particular mental functions were carried out by specific parts of the brain. An alternative view argued that large parts of the brain were equally involved in all mental activities—with no specific functions located within particular parts of the brain. If consciousness is an indivisible whole, the bulk of the brain must function as a single "cerebral organ" controlling all the mental processes.

Gradually the view that there were specialized mental faculties, each with a material substrate in a particular region of the brain, became dominant. The brain was regarded as an aggregate of many organs, each of which embodied one particular faculty (Zola-Morgan 1995). (We mentioned this briefly in an earlier chapter in our discussions of phrenology.) Although the relationship between mental faculties and brain areas was not well understood, researchers began to realize the value of studying this relationship from two directions—the so-called bottom-up approach beginning with neural activity that gradually gives rise to conscious experience, and the top-down approach that begins with conscious experience and expectations that guide and control the activity of specific neural networks.

## THE BOTTOM-UP APPROACH TO THE BRAIN-MIND ISSUE

Historically the bottom-up approach contributed much of the early knowledge of brain-mind links, because the techniques then available made it the most scientifically feasible. Only with the remarkable recent increases in the diversity and sophistication of techniques for imaging and monitoring brain activity online have the top-down approaches begun to bear so much fruit.

With the advent of more scientific approaches to the study of mind-body relations, it gradually became clear that the mind was not a separate entity that was contained within the brain, but rather was a functional property of the brain itself. This was most clearly demonstrated in research on language, perception, and movement.

In the early 1800s French physiologist Jean-Pierre Flourens found that removing an animal's cerebral hemispheres completely erased that animal's perception of the outside world (Flourens 1824). Soon afterward, Marc Dax argued that human speech disorders were linked to damage in the left hemisphere of the brain (Dax 1865). This view was reinforced by Paul Broca, who reported the case of a patient who stopped speaking when pressure was applied to the left frontal lobe of his brain, and of another patient who lost the ability to speak after damage to the left hemisphere (Broca 1861a, 1861b). The localization of language function was further clarified by Carl Wernicke's evidence that different aspects of language and speech were associated with different parts of the brain's left hemisphere (Wernicke 1874). As an interesting side note, at no time did the phrenologists Franz Gall (1835a) and Johann Spurzheim (1827) discuss the possibility that each cerebral hemisphere might have specialized functions, and this despite their preoccupation with localized brain centers.

During the 1870s, a period that Ivan Pavlov described as "an out-standing period in the physiology of the nervous system" (Pavlov 1951, 202), anatomist Gustav Fritsch and psychiatrist Eduard Hitzig carried out an important series of experiments. They demonstrated for the first time that applying electric current to stimulate certain areas of a dog's cerebral cortex—which they called "motor centers"—produced contractions of specific muscles in the dog's body (Fritsch and Hitzig 1870). Next, they damaged small areas in the rear portion of a dog's brain and found that, while the dog could still "see," it could no longer recognize objects.

Over the next half-century, careful research on animals, together with astute clinical observations of humans by neurologists, gradually built up a picture confirming localization of specific functions within the brain. For example, David Ferrier explored the cortex of the brains of monkeys by carefully stimulating and destroying small areas of brain tissue. He showed that the sense of smell depends on a region at the tip of the temporal lobe, the so-called uncinate gyrus (Ferrier 1876). This finding connected with Hughlings Jackson's observations that, in humans, hallucinations of smell often accompanied epileptic seizures that arose from tumors in this part of the brain (Jackson 1899).

Opposing schools of thought continued the debate about the localization of mental functions well into the twentieth century. One group, led by Karl Lashley (1929), promoted the earlier view that mental activity is a single, indivisible phenomenon—the function of the whole brain working as a single entity, under principles Lashley called "mass action" and "equipotentiality." Another group sought to relate each mental process to a particular area within the brain, thus regarding the brain as an aggregate of separate organs. As evidence gradually accumulated about very specific links between mind and brain, this second approach won the day. But in a more general sense, the debate continues. The research of some neuroscientists focuses on exploring examples of tightly constrained localization of function within tiny areas of the brain, while others focus on neural networks, studying the almost unbelievably complex interconnections and interactions between distant parts of the brain, working in unison to perceive the world and act on it.

It is quite remarkable that, in the space of less than a hundred years, the assumption that brain events and mind events were not related had changed to a recognition of clear links between brain areas and language, as well as other mental functions. Although not yet as clear, there were also hints of links between specific areas of the brain and personality, social behavior, and ethical decision-making (see Jeeves 2017).

## The Accelerating Pace of Neuroscience Research

The rapid development of cognitive neuroscience in recent years is generally attributed to the convergence of three previously unrelated areas of scientific endeavor: experimental psychology, comparative neuropsychology, and brain-imaging techniques. The cognitive revolution within experimental psychology freed it from earlier narrowly circumscribed behaviorist approaches to the understanding of mind and behavior. Now psychologists could talk freely about mental events and not simply about stimulus-response contingencies. Also, the development of new experimental techniques enabled cognitive psychologists to divide psychological processes into their component parts. For example, the idea of memory as a single process gave way to multiple memory processes such as short-term memory, working memory, and long-term memory.

In comparative neuropsychology, techniques found useful in studying human cognition were adapted and applied to the study of nonhuman animals. Exciting new findings came from studies of memory and visual perception. For example, the pioneering experiments of David Hubel and Torsten Wiesel—for which they won a Nobel Prize—showed how individual neurons in a cat's visual system are organized in a way that allowed the perception of complex objects (Hubel and Wiesel 1962). Other researchers used single-cell recording techniques to study the neural underpinnings of perception in awake and alert monkeys (Perrett et al. 1984). At the same time, there were exciting developments in brain-imaging techniques, notably positron emission tomography (PET scans), nuclear magnetic resonance (MRI scans), and functional nuclear magnetic resonance imaging (fMRI). These new methods made possible the monitoring of brain activity while people with intact brains performed specific mental tasks.

## Tightening the Links between Mind and Brain

While the contribution from neurophysiology to neuropsychology remains important, a clear distinction can be made between the primary question for the physiologist of "How does the nervous system work?" and that of the neuropsychologist, "How does the working of the nervous system produce behavior, including cognitive processes?" Anatomical and physiological work related to neuropsychology have both thus contributed to intense interest in differences among the cerebral lobes, among the various distinct regions within each lobe, and between the two cerebral hemispheres.

In the mid–twentieth century, research on brain-damaged humans was combined with brain lesioning studies on animals and generally interpreted in a behaviorist framework dominant in North American psychology at the time. This gave rise to what Roger Sperry described as "the traditional micro-determinism of the materialist-behaviorist era emphasizing causal control from below upward" (Sperry 1987, 165). Data from the bottom-up approach, which supported localization and specificity of function, could readily be fitted into such a materialist-behaviorist interpretation. However, more recent evidence became problematic because it raised the possibility of *plasticity* (changes in brain function as a result of experience) and suggested a major causal role for top-down cognitive processes.

According to Roger Sperry, this new evidence produced a shift in the scientific status and treatment of conscious experience—a shift that would have far-ranging philosophic and scientific implications. He argued, "The mind has been restored to the brain of experimental science" (Sperry 1987, 166), and claimed that mental events are "different from, more than, and not reducible to" the brain events from which they emerge (Sperry 1987, 164). However, recognizing that cognitive processes cannot be "reduced to" brain activity does not entail any restriction on neuropsychological research aimed at localizing functions such as memory, speech, and decision-making to specific regions or systems within the brain. The question of whether and to what extent specific mental functions can be localized in particular parts of the brain remains an enduring issue in neuropsychology.

Sperry's experiments with the so-called split-brain patients gave the greatest impetus to the research on the distinctive roles of the two cerebral hemispheres. This led to rapid advances in understanding how cognitive functions were localized differently in the two cerebral hemispheres and how specific mental processes or even component parts of those processes were tightly linked to a single region or system in the brain. Within those regions, moreover, there often emerged a further specificity, indicating that certain columns of cells or small clusters of neurons are involved when a particular aspect of a task is being performed. We give some illustrative examples below.

## SOME EXAMPLES OF LOCALIZATION OF FUNCTION

### Localizing Mechanisms of Perception and Cognition: What's in a Face?

Over the past fifty years there were occasional reports in the neurological literature of patients who, having suffered strokes, said that they could no longer recognize individual human faces, including their own. They could recognize objects, animals, or houses, but not faces. With the advent of brain-scanning techniques, it became possible to identify the specific areas of the brain that, when damaged, seemed to result in problems with face processing.

There was already ample evidence of how visual signals arising from the face provided an abundance of social information about an individual's gender, age, familiarity, emotional state, and potentially their intentions and their mental state. Neuroethologists, having studied the information gained by interpersonal face perception across many species, pointed out that the primate face has evolved an elaborate system of facial musculature that helps in producing expressive facial movements. What soon became clear was that the eyes and the direction of gaze were especially important. The eyes had, of course, long held a special interest to humans. They were said to be "the window of the soul," and normally are one of the first points of contact between infants and their mothers.

Our fascination with the face, especially the eyes, was nicely demonstrated by researchers at Newcastle University in the United Kingdom. In a staff room where people made their coffee, the researchers put up an image of a pair of eyes next to the "honesty box" in which coffee drinkers were expected to contribute toward the cost of coffee supplies. Although none of the forty-eight staff members were aware that an experiment was being conducted, the results showed that the amount contributed to the honesty box increased during weeks when the watching eyes were present, as compared to a different image of flowers (Bateson, Nettle, and Roberts 2006).

Following up an observation by Charles Gross (Bruce, Desimone, and Gross 1981), David Perrett used single-cell recording techniques to show that certain cells in the temporal lobe of monkeys' brains responded selectively to the sight of human faces (Perrett, Rolls, and Caan 1982). Even more surprising, the neuronal response was not affected by changing the size or orientation of the human faces. However, if the individual features of the faces were scrambled—making the image less "facelike"—the responses of the cells diminished.

This finding of a "domain-specific" neural mechanism dedicated to face perception was then replicated by Nancy Kanwisher at MIT (Kanwisher 2000). Extending the evidence from single-cell recording such as used in Perrett's initial studies, Kanwisher used functional magnetic resonance imaging (fMRI) to identify the *fusiform face area* (FFA) of the brain as the place where the specific facial recognition mechanisms are located.

Although there is clear localization of face perception in the brain, Kanwisher also noted that *that does not rule out the need for plasticity*. She referred to evidence for the development of face perception in studies of children born with dense bilateral cataracts. These children have no pattern vision until their cataracts are surgically corrected, which usually happens between two and six months after birth. Successful surgery typically allows pattern vision to develop, but even as adults, these individuals have impaired face perception. Pattern vision in the first few months of life appears necessary for the development of face processing as an adult. Kanwisher concluded an extensive review of the literature with these words: "Evidence indicates important roles for both *genetic factors* and *specific early experience* in the construction of the Fusiform Face Area" (Kanwisher and Yovel 2006, 2123, emphasis added). Clearly, understanding the nature and extent of localization of function in the brain is a complex and ongoing enterprise—something that must be kept in mind when reading media reports giving overly simplistic accounts of localization of specific mental processes within the brain.

Thus, a large amount of evidence has accumulated that demonstrates incontrovertibly the existence of a small region in the primate brain specialized for face perception. But we don't want to leave a false impression that all mental functions are localized to this degree in the "hardware" of brain tissue. Humans and other primates have brains that are powerful general-purpose computers and can be put to work in new ways on novel problems. For example, Christine Keysers and David Perrett make the point that most primates are social creatures who need to understand the actions of others in order to survive. They propose a model of how the monkey brain can recruit three small brain regions specialized for perceiving body motions (such as grasping or dropping an object), and combine their outputs to understand the actions of others by associating them with self-produced actions. They note that this system appears also to exist in humans, and thus their model can provide a framework for understanding the development of human social perception (Keysers and Perrett 2004).

Most of the evidence we have just described comes from neuroscience research investigating how damage to specific brain regions affects face perception. But as was emphasized in the book *Mind Fields* (Jeeves 1994), research on mind-brain links can take several different forms. The contributions of cognitive psychologists to the study of face perception by modeling face recognition processes in terms of modules and dual processing routes represent a different level of analysis that supplements and complements the neuroscience findings.

A recent special issue of the *Quarterly Journal of Experimental Psychology*, devoted to developmental prosopagnosia (sometimes called "face-blindness"), beautifully illustrates how diverse approaches may converge to give a better understanding of the neural bases and psychological processes involved in face perception. The special issue opened with this paragraph:

> Over the last 20 years much attention in the field of face perception had been directed towards the study of developmental prosopagnosia (DP), with some authors investigating the behavioral characteristics of the condition, and many others using these individuals to further our theoretical understanding of the typical face processing system. It is broadly agreed that the term "DP" refers to people who have failed to develop the ability to recognize faces in the absence of neurological illness or injury, yet more precise terminology in relation to potential subtypes of the population are yet to be confirmed. (Bate and Tree 2017, 193)

This is a reminder that difficulties in face perception may occur in the absence of specific brain damage, perhaps as a result of the slowing of developmental processes and/or of reduced efficiency in links between different brain regions.

The speed of ongoing research into face perception over the last twenty-five years provides an excellent example of the point made by Nobel laureate Venki Ramakrishnan, the current president of the Royal Society, which we mentioned earlier:

> The court of science never passes a final judgment but constantly re-evaluates the evidence to arrive at our current understanding. . . . The evidence will win out in the end. Science . . . is not perfect, but science is still our best bet for understanding the world around us and for improving our lives. (Ramakrishnan 2016, 26)

This recognition of the steady advancement of science, including psychology, reminds us of the dangers highlighted in a previous chapter of seeking for concordisms between today's science and the current understanding of scripture by the best biblical scholars. Both realms of knowledge continue to advance, and these advances must be given due recognition. We cannot stop the clock.

These advances in one specialized topic in neuropsychology serve further to underline how misguided it would have been to seek to "integrate" in Jeeves (1994) what was then known about face perception studies with some accounts of human social perception supposedly derived from our Christian beliefs. This would have been as misguided as it would have been to try to demonstrate a concordism between a biblical account of "knowing Jesus" and a psychological model of recognizing the face of Jesus.

## LOCALIZING CHANGES IN BEHAVIOR AND PERSONALITY

In the year 2000, a male schoolteacher began visiting pornographic websites that featured images of children and adolescents. In his own words, "he could not stop himself doing this." When he started making sexual advances to his stepdaughter, his wife called the police. He was arrested for child molestation. He was convicted and underwent a twelve-step rehabilitation program for sexual addicts. The day before his sentencing, he voluntarily went to the hospital emergency room with a severe headache. He was distraught and contemplating suicide. The medical staff who examined him said that "he was totally unable to control his impulses" and "he had propositioned the nurses."

An MRI scan was taken of his brain and revealed an egg-sized tumor pressing on his right frontal lobe. After the frontal lobe tumor was removed, his lewd behavior and pedophilia faded away (Burns and Swerdlow 2003). A year later the tumor partially grew back, and the man started once again to collect pornography. A further operation was undertaken to remove the regrowing tumor, and his urges again subsided.

There was widespread comment on this case. A neurologist said that "he saw people with brain tumors who would lie, damage property, and in extreme rare cases, commit murder." He further commented, "The individuals simply lose the ability to control impulses or anticipate the consequences of choices." A psychiatrist specializing in behavioral changes associated with brain disorders and who had studied the way in which brain tumors can affect a person's behavior commented, "This tells us something about being human, doesn't it? . . . If one's actions are governed by how well the brain is working, does it mean we have less free will than we think?" (Kahn 2003).

These specialists know that human behavior is governed by complex interactions in the brain. Many neuroscientists believe that so-called executive functions—that is, higher-level decisions with major consequences—are dependent upon the intact functioning of systems within or linked to the frontal lobes, which are regarded as the most highly evolved area of the brain. Tumors in this area can squeeze enough blood from the region to effectively put it to sleep, thus dulling someone's judgment in a way similar to drinking too much alcohol. However, only in very rare cases will the tumor turn the person to violence or deviant behavior.

The dramatic changes and then reversal in the behavior of the school-teacher described above provide a vivid illustration of how *our moral behavior is embodied in our physical makeup*. Similar dramatic effects have been on record for a long time. Every student of neurology and neuropsychology has heard about Phineas Gage, the railroad worker who damaged his frontal lobe in an accident and whose behavior changed for the worse—from a reliable, industrious pillar of society to an irritable, profane, dishonest man—at least for a period of time after his injury (Van Horn et al. 2012). As Frans de Waal, when commenting on two similar patients studied by Antonio Damasio, wrote:

> It's as if the moral compass of these people has been demagnetized, causing it to spin out of control. What this incident teaches us is that conscience is not some disembodied concept that can be understood only on the basis of culture and religion. (de Waal 1996, 217)

Morality, de Waal claimed, is as firmly grounded in neurobiology as anything else we are or do. However, Alexander McCall Smith, former professor of medical law and expert in bioethics, writing on the impact of neuroscience and insights that it provides, cautioned that

> from both the moral and legal point of view, the fact that a person has acted in a particular way because of a characteristic of the brain or of the genotype, may not exculpate but may then lead us to mitigate blameworthiness to some extent. Returning to the psychopath, or indeed to the pedophile, such persons may behave anti-socially, and may be held responsible for what they do, but they still deserve some pity at least on the grounds that the starting point for their choices is so much more difficult than it is for those who are not affected by such conditions. (McCall Smith 2004, 121)

A recent article titled "How Responsible Are Killers with Brain Damage?" raises a similar question. After examining seventeen cases in which law-abiding citizens had developed a brain lesion and then engaged in criminal behavior, the author wondered, "If their actions were caused by brain damage and a disrupted neural network, were they acting under their own free will? Should they be held morally responsible for their actions and found guilty in a court of law?" (Micah Johnson 2018).

The more we discover about the genetic and neural underpinnings of behavior, the more it should generate compassion, which according to scripture is a Christian hallmark. Here then is another example of advances in science that give new insights into how we may seek to show compassion in all our relationships.

Before we leave this reminder of the close links between the intactness of our brains and our capacity for moral behavior, we should perhaps pause and reflect on the results of a recent study reported by Gesch and his colleagues showing that physical changes that affect behavior may be very small and subtle and not dependent upon widespread damage to neural structures. They reported that boosting the dietary intake of certain vitamins, minerals, and fatty acids reduced the occurrence of antisocial behavior (Gesch et al. 2002). In contrast, reducing the amounts of these supplements led to an increase in antisocial behavior.

## Top-Down Effects: The Accumulating Evidence

### Clues from Neurochemistry

Every aspect of our behavior—our decisions, actions, and subjective experiences—is linked with and dependent upon neurobiological processes. Changes in what we experience and how we feel about those experiences are fundamentally tied to alterations in brain activity—the workings of billions of nerve cells and trillions of ever-changing connections between the cells as they are modulated by the release of tiny packets of chemical messengers called *neurotransmitters*. While we do not fully understand the nature of the relationship between subjective mental phenomena and observable brain processes, advances in neuroscience have made clear an intimate link between the two. For example, one area of neuroscience where the relationship between mental state and neural activity has been studied extensively is that of bodily pain. It is now beyond doubt that one's mental attitude and what one expects to experience has a robust and measurable effect on the degree of pain reported when the body is stimulated. The so-called placebo analgesic effect is a well-known example of this.

A typical pain/placebo experiment might train a participant to block the pain with an infusion of morphine, and then later switch to an infusion of an inert solution such as salt water. If the participant *believes* the solution is morphine, the experienced pain is dramatically reduced (Petrovic et al. 2002). When brain imaging is also carried out on participants who show this robust placebo analgesia effect, the images confirm their subjective reports by showing reduced activity in brain areas normally activated by painful stimuli, while other brain areas activated by the powerful painkiller morphine tend to show increased activity. A further discovery is that the placebo analgesic effect can be reversed by a molecule that blocks the brain's receptor for opioids.

It is not surprising that placebo analgesia has become an important component in pain management, even though the basic mechanisms are still poorly understood. From what we do understand, it appears that placebo analgesia involves endogenous opioid systems in the brain as well as higher-order cognitive networks. It has been shown that brain regions called the *rostral anterior cingulate cortex* and the *brainstem* are involved in opioid analgesia, and it is likely that these structures play a similar role in placebo analgesia. A study using positron emission tomography (PET) to study the brain has confirmed that both opioid and placebo analgesia are associated with increased activity in the rostral anterior cingulate cortex (Petrovic et al. 2002). The point to remember: *what we believe and what we think modify the activity of our brain.* The mind-brain connection flows in both directions.

## *Clues about Mental Causation*

Any practicing clinical neurologist can give you examples of how patients suffering from epileptic seizures may, in some cases, develop psychological tricks that enable them to avert a seizure. One patient described how, when she felt a seizure was about to start, she would ask someone to read to her so that she could avert the seizure. Listening intently to music apparently had a similar result. Amazingly, an ordinary perceptual experience combined with the determination to resist the seizure are, as one neurologist has written, "plainly doing business with a rebellious brain." Such deliberate, conscious attempts to influence brain activity would be examples of what the late Nobel laureate neuroscientist Roger Sperry called "emergent top-down control" (Sperry 1987, 165). These he contrasted with the so-called bottom-up effects (brain activity influencing mental events) that we described earlier.

In a dramatic illustration of the rate of progress in cognitive neuroscience, Thomas Grabowski and Antonio Damasio wrote in 1996 that "the imaging of the neural correlates of single and discrete mental events, such as one image or one word, remains the most desirable dream" (Grabowski and Damasio 1996, 14303). A mere four years later, this "neuroscience dream" became reality in research by Kathleen O'Craven and Nancy Kanwisher. Their study (O'Craven and Kanwisher 2000) beautifully illustrated how top-down mental processes can selectively mobilize specific areas and systems in the brain. They asked their subjects either *to look* at pictures of famous faces or familiar places, or *to imagine* them (i.e., form a visual image of each face or geographical location). They showed that imagining faces selectively activated the same area of the brain (the fusiform face area) as when the subjects were seeing the actual pictures of those faces. In contrast, imagining familiar places or looking at pictures of those places both activated the *parahippocampal area*. The experimenters demonstrated that they could, in a sense, "read the minds" of their subjects by observing their brain activity. They could tell whether the subjects were thinking about faces or places by measuring activity in the respective brain areas. Going back to the wistful quotation from Grabowski and Damasio, we note the danger of Christians reacting to perceived threats against traditional models of human nature by proposing things that science "will never" be able to accomplish. It's not wise to bet against scientific progress powered by God-given human intelligence.

## Neural Plasticity

### Plasticity in Sensory Processing

Recent research using newly developed brain-imaging techniques adds to our confidence in the *plasticity* of the brain—the idea that brain activity can be shaped or "rewired" as a result of environmental experiences. Norihiro Sadato and his colleagues studied people who had been born blind and had been trained to become extremely efficient in the use of Braille. The researchers discovered that some areas of the participants' brains normally devoted to vision had been taken over by touch (Sadato et al. 1996). These reports from the studies of humans were subsequently replicated in monkeys using single-cell recording techniques.

Another dramatic and widely reported example of plasticity was found by Eleanor Maguire and her colleagues, who used modern brain-imaging techniques to study the brains of London taxi drivers. It is well known that licensed London taxi drivers are renowned for their extensive and detailed navigation experience and skills. Maguire studied structural MRIs of their brains and compared them with those of matched control participants who did not drive taxis. They reported that the posterior hippocampus of the London taxi drivers was significantly larger than those of control participants. The hippocampal volume also correlated with the amount of time spent as a taxi driver. The researchers concluded, "It seems that there is a capacity for local plastic changes in the structure of the healthy adult human brain in response to environmental demands" (Maguire et al. 2000, 4398). When the hippocampus is used extensively, there are measurable changes in its shape and size, suggesting that a hippocampus capable of good navigational skills is not predetermined exclusively by genes. Even more persuasive was the fact that some of these same taxi drivers showed a reduction in hippocampal size after retirement, when they were no longer exercising their navigational skills to the same extent (Woollett, Spiers, and Maguire 2009).

### Plasticity in Brain Development and Face Perception

At a 1997 meeting convened by US president Bill Clinton and his wife, Hillary, it was suggested that a valuable investment in the future would be to provide "enriched" environments for infants and young children (Clinton and Clinton 1997). In the context of the ongoing nature-nurture debate about whether the mental capacities of young children are innate or learned, this initiative drew support from the nurture side of the debate. Today, however, most researchers in the field have moved beyond such a simple dichotomy. The debate today is much more sophisticated, focusing on issues such as the effects of experience on the way that infants perceive and process objects. Also, recently some have argued for what they call a "constructivist" point of view. This view argues that development is not merely the implementation of a genetic blueprint nor that the infant's brain is a passive slate written on by experience. Instead, it claims that genes and environment *interact* at multiple levels in a constructive manner (Karmiloff-Smith 2009).

Some of the newest developmental research depends on a recently perfected safe and easy method to study the workings of a baby's brain by recording "high-density event-related potentials." Researchers such as Mark Johnson (Johnson et al. 2001) believe that this approach will revolutionize the study of the baby's brain at work. The method involves gently placing a "geodesic" net made up of large number of passive sensors (electrodes) on the baby's head. These sensors on the scalp pick up the natural electrical changes as groups of neurons are activated within the brain. The output from these sensors is processed by a computer to produce a map of active regions across the baby's head.

While most of the neurons of the brain are in place at birth, dramatic increases in *axon* branches (the conducting fibers emerging from the neuron cell bodies) and in the number of *synapses* (the special junctions between neurons that pass signals from one neuron to the next) occur during the first few months of life. Many of the axons are gradually covered with *myelin*—a fatty sheath rather like the plastic covering around household electric wires—that helps to ensure good conduction along the fibers. In most of the regions of the cerebral cortex that have so far been studied, the density of the synapses increases in infants to a point about half again as much as in adults. Following this there is a "neural pruning" phase during which many of the synapses die and disappear, reducing the overall level to the number we observe in adults.

These changes make it clear that brain development is not the passive unfolding of a genetic plan, but very much a dynamic, activity-dependent process. To give an example of this, we can refer once again to research on face processing. It has been known for some time that newborn infants, only a few hours old, tend to turn their head and eyes to look at faces more often than to any other complex pattern that has so far been studied. Mark Johnson (2000) has suggested that this newborn behavior is a reflexlike response controlled by some of the older subcortical parts of the brain. Using the geodesic device described earlier, Johnson and his colleagues have discovered that when you present a face image to an infant, the image often elicits more widespread brain activity in infants than in adults. This suggests that the brains of young infants have "surplus" connections among neurons that are pruned over time, or at least become progressively more focused in their activity. For example, adult brains show different response patterns to pictures of human faces compared to monkey faces. This difference does not occur in babies. Thus, parts of infants' brains seemed to be less finely tuned than the same regions in adults.

Why did early developmental theorists believe that babies' brains were passively shaped by environment experiences? One reason is that, in the first few months of life, infants cannot walk or even reach accurately. But appearances can be deceptive. Johnson and his colleagues have shown that newborn infants are quite capable of directing their eyes toward things that interest them—such as faces. Even their very limited motor control does not prevent them from actively searching out their preferred visual stimulation, and in turn, that stimulation generates and modifies brain circuits in the cerebral cortex. By four months, babies have already made over 3 million eye movements. Thus, Johnson concludes that *infants actively contribute to their own subsequent brain development* (Mark Johnson 2000). The malleability of the infant brain may also turn out to be its salvation. If the specializations of regions of the brain can be modified by experience, there is some hope of finding ways of ensuring that the brain receives the appropriate inputs for optimal development. Plasticity is the name of the game.

A study by Christy Marshuetz and her colleagues (Payer et al. 2006) further demonstrated the plasticity of the adult brain. It built upon the previously mentioned work of O'Craven and Kanwisher (2000), who demonstrated that different areas of the brain become active depending on whether a person is looking at a face or a place. These researchers asked two groups of people, one ages eighteen to twenty-seven years and the other sixty-one to eighty years, to remember three images of faces, and then asked them to decide if another face image was from the original set. Next, they asked the participants to remember three images of houses (geographical scenes) and then decide whether another house image was from the original set. They tracked the neural changes in the participants' brains using functional magnetic resonance imaging (fMRI). In reporting the results, they made the point that it has been known for some time that, as we get older, our neural and cognitive functions become less differentiated. The results of this study indicated that the older adults did indeed show decreased specialization in the "face area" of the brain and in the "place area" of the brain when compared with the younger adults. They also found that older adults showed more activity in the frontal cortex, which the researchers believed was compensation for less differentiation in the visual areas at the back of the brain. They concluded, "Our findings are the first to demonstrate decreased neural specialization in the ventral visual cortex in older adults, along with increased activation in the prefrontal cortex. . . . This underscores the importance of taking into account the connected and network nature of the brain and its function in understanding human neural aging" (Payer et al. 2006, 491). This research provides yet another example of plasticity alongside fixed specialization in the functions of the human brain.

## Making Sense of It All

It is one thing to demonstrate the tight interaction between what is happening at the conscious mental level and what is happening at the physiological level of the brain and the body, but the unanswered question is, how can we most accurately characterize this intimate relationship without making claims or assumptions that have not yet been demonstrated? What is clear is that there is a remarkable *interdependence* between what is occurring at the cognitive level and what is occurring at the physical level. We could perhaps describe this as a relationship of *intrinsic interdependence*, using the word "intrinsic" to mean the natural way the world operates (*Concise Oxford Dictionary:* "intrinsic" = belonging naturally, inherent). Could we perhaps go further than this and say that, based on our present knowledge, it is an *irreducible intrinsic interdependence* (Jeeves 2013, 40), meaning that *we cannot reduce the mental to the physical any more than we can reduce the physical to the mental*? In this sense, there is an important duality to be recognized, but it is a *duality without dualism* (MacKay 1978; Ludwig 1997).

On this point, there has fortunately been somewhat of a convergence of views between neurologists and psychiatrists. This is illustrated by two quotes. The distinguished neurologist Antonio Damasio wrote:

> The distinction between diseases of brain and mind, and between neurological problems and psychological/psychiatric

ones, is an unfortunate cultural inheritance that permeates society and medicine. It reflects a basic ignorance of the relation between brain and mind. (Damasio 1994, 40)

These comments were echoed by a recent past president of the Royal College of Psychiatrists in Britain, Robert Kendell, when he wrote:

Not only is the distinction between mental and physical illfounded and incompatible with contemporary understanding of disease, it is also damaging for the long-term interests of patients themselves. (Kendell 2001, 492)

In light of recent neuroscience advances, perhaps the mind-brain issue, once thought to raise deep theological questions sparking conflict between science and religion, may become less of a problem for knowledgeable Christians in the future.

## Concluding Cautionary Comments

In the concluding chapter to the book *The New Brain Sciences: Perils and Prospects*, Sir Dai Rees and Barbro Westerholm emphasized:

These insights [from neuroscience] can only heighten the sense of mystery at the workings of the human mind rather than encourage any idea that the remorseless march of the new brain sciences will soon arrive at a "total explanation" or "final synthesis" of them. On the contrary, the picture seems to expand and become ever more elaborate. (Rees and Westerholm 2004, 267)

If we believe that all truth comes from God, we can, as Christians who are scientists, be quite relaxed about the wonders uncovered every day by research on human nature. What we already know will seem so small in comparison to what will be revealed in the coming decades. The new discoveries will add even further to our conviction that we are indeed "fearfully and wonderfully made" (Psalm 139:14). In the meantime, we are left with many fascinating and challenging issues at the interfaces of our disciplines.

CHAPTER 7

# Insights from Neuropsychology
# about Spirituality

IN THIS CHAPTER and the next, we provide examples that illustrate
how, by bringing together advances in neuropsychology with advances
in biblical scholarship, we may discover new insights into our faith—
insights that may enrich the quality of Christian life for individuals
and groups. The focus of this chapter is on how neuropsychology and
biblical scholarship mutually enrich our understanding of spirituality.
Specifically, we shall consider

- Embodied spirituality
- Embedded spirituality
- Impaired spirituality

## SOME HISTORICAL PERSPECTIVE

Despite what some Christian books and television programs assert,
living the Christian life is not a bed of roses. Only someone totally
ignorant of the history of the Christian church could take such a view.
Careful historical research has documented in detail the struggles of
some great men and women of the faith, revealing a close connection
between psychological experiences and faith. As psychiatrist Gaius
Davies has written:

The question of how temperament and faith are connected is, of course, brought to the fore in every conversion experience. We cannot understand Methodism without knowing something of how John and Charles Wesley found faith and assurance in 1738. Both found Martin Luther a great catalyst: John through Luther's work on Romans, Charles through Luther's commentary on St Paul's letter to the Galatians. (Davies 2001, 11)

Gaius Davies's detailed studies of well-known Christian figures such as Martin Luther, John Bunyan, John Wesley, William Cowper, Gerard Manley Hopkins, Lord Shaftesbury, and Christina Rossetti made clear that these outstanding leaders who contributed so much also suffered a great deal from anxiety and depression. As John Stott wrote in a foreword, "What I especially admire about Gaius Davies' book is his honesty and realism. He offers no glib remedies. It tells us the truth, that some of God's heroes and heroines have been eccentric and neurotic, and have suffered repeated breakdowns" (Davies 2001, 6). What is particularly relevant in the present context is Davies's argument—based on recent discoveries about the biological roots of behavior and the impact of psychotropic drugs—that the stories of some of these heroes and heroines of the faith may have been quite different had they lived today. Consider a few examples by way of illustration.

Writing of Martin Luther, Davies said, "I believe there was a marked physical and constitutional element in Luther's tendency to depression. I do not see clear evidence that he was ever manic, elated or ill because of an upward mood swing. However, he might now be diagnosed as a cyclothymic personality, with many mood swings which, though significant, were never such as to cause psychosis" (Davies 2001, 45). Davies documents the evidence that in 1527 Luther was suffering both physically and with depression. Although only age forty-four, he collapsed physically and was expected to die. Luther himself later saw this episode of illness as partly physical and partly psychological (Luther 1955, 115–17).

Of John Bunyan, Davies wrote:

> In *Grace Abounding*, Bunyan describes his severe anxiety, and
> how it often drove him to despair. However, he was not a monk
> with hours to spend in the confessional. Unlike Luther, he was
> much more alone with the Bible, and struggled with difficult
> texts with little help, even allowing for John Gifford and the
> few books he could obtain. (Davies 2001, 65)

Davies went on to say about Bunyan, "There could be no clearer example of the severe obsessive-compulsive disorder at work. There are other
examples of the same urge to blaspheme. He would want to hold his hand
over his mouth, or plunge his head into a dung heap rather than give in
to the impulse" (Davies 2001, 65–66). Later he continued, "Bunyan was
remarkable because he suffered so severely from obsessions; he would
nowadays be diagnosed as being in need of treatment" (Davies 2001,
66). Davies not only described the symptoms Bunyan suffered and the
way Bunyan obtained relief, but he also showed that great strength may
result from overcoming a serious psychological disorder. One can only
speculate how much John Bunyan would have benefited by the judicious
use of some of today's effective psychotropic drugs.

Writing of William Cowper, Davies said:

> Cowper wanted to break the conspiracy of silence about
> depression. . . . Cowper was depressed first at the age of
> twenty-one, and from time to time thereafter for the next ten
> years. At thirty-one he had his first catastrophic psychotic
> breakdown, and at the time of his recovery from it he became
> a Christian. He was to have four more depressive illnesses
> before he died at sixty-eight: in between these times he was
> often amazingly productive as a letter writer and poet. (Davies
> 2001, 93–94)

Near the end of his chapter on William Cowper, Davies wrote, "Why was he not healed? Why should he have suffered six serious depressive breakdowns, several suicide attempts and endured so much mental pain? It is part of the larger mystery of suffering, and there cannot be a final answer. But good came out of the apparent evil of his distress" (Davies 2001, 118). Again, in the instance of William Cowper we can but speculate what a difference it might have made had some of today's antidepressant medicines been available.

Biblical scholar J. B. Phillips was described by Davies as someone who had "a special genius for translating and communicating." Despite all that Phillips achieved, it is clear from his autobiography *The Price of Success* that he suffered from recurrent depression. In his sympathetic description of Phillips's many struggles, Davies wrote:

> I find it sad that he was not helped by medication. Nowadays such non-psychiatric forms of treatment as the use of beta-blockers may alleviate the anxiety by stopping the excessive effects of adrenaline on the body: they do this without sedation and without any risk of dependence. By the same token, there are nowadays many forms of antidepressant which might help someone who suffered as much as Phillips. (Davies 2001, 320)

In the final pages of his book, Davies drew together these threads under the heading "Personality and Temperament," writing:

> My aim in the previous studies in this book has been not only to avoid speculation, but also to draw some conclusions from the facts reviewed. I have tried to show how Martin Luther and John Bunyan were perfectionists who suffered a good deal in their youth from obsessional symptoms. The careful reader will have noted a number of the others—J. B. Phillips, Christina Rossetti, Amy Carmichael and Gerard Manley Hopkins were also people with marked traits of obsessionality. (Davies 2001, 372)

Davies concluded with some profound advice:

> A kind of miracle happens when our heroes find that the water
> of life, as Christ's gospel has become for them, is turned into a
> heady wine of doctrinal delight. As C. S. Lewis put it, the heart
> sings unbidden not with the devotional book but in reading
> some Christian treatise which speaks to the mind. This may
> of course become addictive, and systematic theology may be
> taken as a substitute for the real thing, life in Christ. Perhaps
> we live today in a time when we fly from the mind, to urge
> Christians to seek experience, ecstatic phenomena and have
> their feelings stirred up. Bunyan saw it all with the Ranters
> and the early Quakers; Luther with the ecstatic prophets of his
> day. There are, however, hopeful signs of a return to a more
> balanced life where heart and head act together in a better
> balance. (Davies 2001, 382)

The take-home message from Gaius Davies is that our spirituality is related to and influenced by the functioning of our brains and bodies—a message that is underlined by the neuroscience research described in this chapter.

## What Is "Spirituality"?

In an extensive and detailed review of the varied ways in which the word "spirituality" is used today, Victor Copan has helpfully and concisely summarized the contemporary situation:

> For more than two decades now, there has been an increasing
> fascination about all things spiritual in the Western world—
> and it is everywhere. The term *spirituality* is a buzz-word in
> popular culture. . . . But what do we mean when we use the
> word "spirituality"? . . . What is a biblical understanding of

spirituality and spiritual formation? . . . It is one of the astonishing developments of the last decades that spirituality has made a strong comeback after years of being out of vogue. Do a Google search on the term *spirituality*, and you'll get over 141,000,000 hits. (Copan 2016, 1)

As yet another example of the widespread fascination with "spiritual experiences," a recent article on psychedelic drugs in *The Psychologist*, the professional journal of the British Psychological Society, stated:

Is there anything psychedelic drugs can't do? A recent wave of scientific scrutiny has revealed that they can elicit "spiritual" experiences, alleviate end-of-life angst, and perhaps treat depression—and they might achieve at least some of all this by "heightening consciousness." (Young 2017, 17)

Others have voiced concerns with and misunderstandings about "spirituality." For example, Donald Carson wrote:

"Spirituality" has become such an ill-defined, amorphous entity that it covers all kinds of phenomena that an earlier generation of Christians . . . would have dismissed as error, or even as "paganism" or "heathenism." . . . It is becoming exceedingly difficult to exclude absolutely anything from the purview [domain] of spirituality, provided that there is some sort of experiential component in the mix. (Carson 1994, 381, 384)

In contrast, for N. T. Wright, the central aspects of Christian spirituality are anything but vague. For him, they include formative spiritual experiences such as new birth and baptism, prayer, reading of scripture, participation in the Eucharist, and the capacity to give and respond to love. The concrete, substantive nature of true spirituality has been further underlined by Miroslav Volf: "Some people like to keep their spirituality and theology neatly separated, the way someone may want to have the main dish and the salad served separately during a meal. I don't. Spirituality that's not theological will grope in the darkness, and theology that is not spiritual will be emptied of its most important content" (Volf 2005, 236).

## UNDERSTANDING SPIRITUALITY: DUALISM VS. MONISM

### Listening to Biblical Scholars, Theologians, and Neuropsychologists

We believe that spirituality is not, as was contended for centuries, something to be located "in an immaterial immortal soul," hidden somewhere in our heads. Nor is it something securely protected from the experiences of living in the world, which can include the effects of brain damage, psychological disorders, or the illnesses of old age. More than a decade ago, N. T. Wright wrote, "Despite what many people think, the Bible does not envisage human beings as split-level creatures (with, say, a distinct body and soul) but as complex, integrated wholes" (N. T. Wright 2004). In a similar vein, Anthony Thiselton, another leading contemporary theologian, has written, "Paul believes in what today we call the psychosomatic unity of the whole person" (Thiselton 2009, 69).

It is true that for centuries, traditional Christian thought saw the body as frail and finite, but the soul as immortal and impervious to damage. Was this dualistic body-soul distinction caused by relying too much on a Platonic "Greek" view of persons rather than a more biblical "Hebrew" view? Today, biblical scholars recognize it as an oversimplification to claim that there was a dramatic gap between Hebrew thought (which affirms some form of psychosomatic unity—what today we call our *psychobiological unity*) and Greek thought (which affirms some form of *dualism*—a separate soul and body). Greek thought was in fact much more varied on the nature of the soul than Plato's views would suggest. As one scholar has put it, "There was no singular conception of the soul among the Greeks, and the body-soul relationship was variously assessed among philosophers and physicians in the Hellenistic period" (Green 2008, 53).

As noted by N. T. Wright (2004), the controversy across the centuries about the relative merits of dualism and monism has not yet reached a definitive conclusion. At present, there is no consensus on the issues surrounding dualism and the soul, as is demonstrated by the 2016 book *Neuroscience and the Soul: The Human Person in Philosophy, Science, and Theology* (Crisp, Porter, and Ten Elshof 2016). In this book, a group of distinguished philosophers and theologians debate the issues and unfortunately reach no agreement. In our view, this issue is at the heart of many contemporary debates about the human person. We turn to that issue now and attempt to show that our "spirituality" is firmly embodied in our physical makeup.

Our approach here is to examine how the new evidence from scientific research may provide fresh insights about spirituality, while avoiding the danger of being so dazzled by discoveries in neuroscience that if an explanation can be given which refers to neuroscience, then that explanation is accorded more weight and validity than other explanations (Fernandez-Duque et al. 2015). As active researchers, we are very careful never to forget the ever-changing scientific landscape to which we are contributing, while noting those areas of knowledge where converging evidence leads us to have increasing confidence in the results—especially when the relevant advances in biblical scholarship point to the same conclusions. As scientists, it would be impertinent of us to claim any special competence to participate in the ongoing biblical/theological debate about the scriptural view of personhood. However, because what one believes about the soul is relevant to discussions of embodied spirituality, we offer our current understanding of what theologians and biblical scholars are saying.

Within mainline Catholic and Protestant traditions there are strong affirmations of belief in an immaterial and immortal soul as distinct from the physical body. For example, the official *Catechism of the Catholic Church* states, "The Church teaches that every spiritual soul is created immediately by God—it is not 'produced' by the parents—and also that it is immortal: it does not perish when it separates from the body at death, and it will be reunited with the body at the final Resurrection" (Catholic Church 2012, 1.2.1.6.366). John Calvin wrote something similar:

> Moreover, there can be no question that man consists of a body and a soul; meaning by soul, an immortal though created essence, which is his nobler part. Sometimes he is called a spirit. . . . And Christ, in commending his spirit to the Father (Luke 23:46), and Stephen his to Christ (Acts 7:59), simply mean, that when the soul is freed from the prison-house of the body, God becomes its perpetual keeper. (Calvin 1960, 1.15.2)

Body-soul dualism is a widely held belief not only among Christians but also in most other religious traditions. It is so deeply rooted in human culture that some, including developmental psychologist Paul Bloom, believe that young children are "natural-born dualists," biologically predisposed to perceive the world in dualistic terms (Bloom 2004). In spite of the popularity of dualism, we believe that the evidence from neuroscience and contemporary biblical scholarship offers fresh insights that encourage us to consider reinterpreting some of the ancient texts and returning to a more holistic view of the human person.

For centuries, the words "soul" and "mind" were used interchangeably. Within the Christian context, the traditional proof text for the existence and importance of the soul is found at the beginning of the Bible in Genesis 2:7: "And the LORD God formed man of the dust of the ground, and breathed into his nostrils the breath of life; and man became a living soul" (KJV). This, for centuries, led to the view that the soul departs the body at death, lives in the spiritual domain, and is reunited with the body at the resurrection of the dead on the last day. Because the soul was thought to be the most important and most distinctive part of the person, it became vital to defend the status of the soul when defining what it means to be human.

However, such a view is widely challenged today by biblical scholars, who see human uniqueness arising not from our possession of an immortal soul representing the "image of God," but rather from our calling to enter into the personal relationship offered to us by our Creator. Thus, Vladimir Lossky writes, "Creation in the image and likeness of God implies the idea of participation in the divine Being, of communion with God" (Lossky 1991, 118). Anthony Thiselton, one of today's leading biblical scholars and an expert on hermeneutics, claims that "to be made in, or called to, the image of God is to represent God" (Thiselton 2015, 477). Another biblical scholar, N. T. Wright, has written, "I have been arguing for some time: that image of God was not in Genesis 1 intended to refer to some characteristic or special ability or trait of humans but rather a vocation. The vocation in question is that humans were designed by the Creator to have a special role in his governance of the world. Eventually it comes round to using the royal priesthood language which I think is absolutely central" (N. T. Wright 2011, personal communication).

Contemporary biblical scholarship is challenging the traditional dualistic view of the soul as separate from the body. Referring to the Hebrew word *nephesh*, which is frequently translated as "soul" in older versions of the Bible, Old Testament scholar Lawson Stone wrote:

> We are not to imagine Adam's reception of some intangible personal essence makes him human, distinct from the animals, and eligible for everlasting life. The *nephesh* is not a possession, nor is it a component of Adam's nature. . . . The pile of dust, upon being inspirited by divine breath actually became a living *nephesh*. The term "living *nephesh*" then denotes the

totality of Adam's being. Adam does not "have" a *nephesh*, he is a "living *nephesh*." (Stone 2004, 49)

Writing about the use of the word "soul" in Genesis, New Testament scholar Joel Green stated, "The term in question, *nephesh*, is used only a few verses earlier with reference to 'every beast of the earth,' 'every bird in the air' and 'everything that creeps on the earth'—that is, to everything 'in which there is life.' . . . This demonstrates that the 'soul' is not for the Genesis story a unique characteristic of the human person" (Green 2008, 64). Moreover, cognizant of the rapid advances in genetics, evolutionary theory, and neuroscience, Green commented, "The neuroscientific tightening of the mind-brain link renders more and more improbable the need for a soul as an ontologically separate entity, and the reality that human self-consciousness is neurobiologically generated is becoming an inescapable datum for Christian anthropology" (Green 2004, 179).

Rod Scott and Raymond Phinney have written about the relationship between body and soul on the basis of insights drawn from human development and neurobiology. In a section headed "Insights from Scripture," they write:

> We wish to demonstrate that the Bible does not provide an explicit anthropology that supports a dualistic perspective. We agree with theologian Joel Green who has written in support of a monistic perspective of personhood. (Scott and Phinney 2012, 91)

However, in order to avoid the impression of selectively citing antidualist authors, Scott and Phinney refer to the writings of theologian John Cooper, who holds a position he calls "holistic dualism." Even Cooper concedes that in biblical word studies, "the variety and interchangeability of terminology simply do not provide a footing for a clearly dualistic reading" (Scott and Phinney 2012, 91).

## Embodied Spirituality

### Neurotheology

Anyone unfamiliar with neurological observations over more than a century may be forgiven for thinking that what is today labeled "neurotheology" is something recent. But awareness of a link between religion and the brain is not new. The link between spirituality and brain processes has a long, well-documented history. There are, for example, both ancient and modern accounts of mystical experiences associated with the use of hallucinogenic drugs, as well as the long-standing association in clinical neurology between certain forms of epilepsy and exceptional religiosity, which we describe later.

As in the case of the top-down studies described in the previous chapter, it has been the rapid sophistication in brain-imaging techniques applied to the study of religious behavior that is responsible for progress in this field, sometimes described variously as *neurotheology* or *theoneurology*. There are potential benefits from good research in neurotheology, such as giving a better understanding of some bizarre manifestations of religiosity and the possibility of bringing relief to those individuals through tailored and targeted psychotropic drugs. But in the context of this book, research in neurotheology is relevant *because it supports the idea that spirituality is firmly embodied in our biological makeup.*

Recent neuroscience research using functional brain-imaging is making it increasingly clear that our religious and spiritual experiences, like all our experiences, are grounded in neural networks within our brains. We shall consider a few examples that illustrate this theme and provide fresh insights into how "wonderfully made" we are (Psalm 139:14). However, these situations also highlight the difficulty of classifying any particular period of strong emotion as a "spiritual experience" rather than an ordinary mood swing. For example, Lewis Smedes, formerly of Fuller Seminary, wrote in very personal terms about his Christian pilgrimage and the way in which modern brain-altering drugs transformed and restored his experienced "spirituality," as indicated in this passage:

> Then God came back. He broke through my terror and said: "I will never let you fall. I will always hold you up." I felt as if I had been lifted from a black pit straight up into joy. . . . I have not been neurotically depressed since that day, though I must, to be honest, tell you that God also comes to me each morning and offers me a 20 mg. capsule of Prozac. . . . I swallow every capsule with gratitude to God. (Smedes 2003, 133)

### Hallucinogenic Drugs and Religious Experience

In the previous chapter, we referred to what we called "clues from neurochemistry" in our review of so-called top-down effects. As a concrete example in the religious domain, some ancient religious rituals used plants to facilitate ecstatic and mystical states: mushrooms (used by the Aztecs), peyote cactus (used by the Huichol of Mexico), and ayahuasca (used by native groups in the Amazon basin), as well as substances derived from water lilies, mandrake, opium poppies, morning glories, and marijuana plants. Since these drugs produce their effects by altering brain activity, research on these drugs can reveal brain mechanisms relevant to the experiences that people often describe as religious.

Chemicals that create the hallucinogenic experience have been found to activate the serotonin network of the brain via the $5\text{-HT}_{2A}$ receptor, which in turn affects many complex brain systems. Although the relationship between the sites and mechanisms of action and the subjective experiences elicited by the drugs is not fully understood, here is what we know at present (D. Johnson 2017):

▸ These drugs increase the release of *glutamate*, the main excitatory neurotransmitter. This increases neural activity across the entire cerebral cortex.

► Neurons in a midbrain area called the *raphé nucleus* have been found to reduce their firing rate under the influence of these drugs. This change in the activity of *raphé* neurons also occurs in dreaming (REM) sleep, and is thought to be related to diminished vigilance and diminished awareness of external stimuli.

► The thalamus also has many 5-HT$_{2A}$ receptors. Drug-related changes in thalamic activity appear to alter the flow of sensory information to the cerebral cortex, creating more sensory noise in cortical information processing, perhaps giving rise to hallucinations (false sensations not related to external events).

► The *locus coeruleus* in the midbrain is also affected by hallucinogenic drugs. This area is involved in the detection of novelty and important external stimuli. Changes in the activity of this nucleus could cause stimuli normally not considered noteworthy to be perceived as significant.

► Hallucinogens also affect the *ventral tegmental area* of the midbrain that projects dopamine-releasing axons to the cortical and subcortical structures. The result of dopamine increase is to mark events as biologically significant, causing memory systems to be activated (T. Lewis 2016).

### Religious Experiences Are Embodied and "Embrained"

Thus, it appears that the *objective*, measurable changes in the brain's serotonin and dopamine systems produced by hallucinogenic drugs influence the *subjective* qualities of psychedelic or religious experiences. Based on what is known about the neurotransmitter systems and the changes produced by the drugs, it is likely that such drugs

> perturb the key brain structures that inform us about our world, tell us when to pay attention, and interpret what is real. Psychedelic drugs activate ancient brain systems that project to all of the forebrain structures that are involved in

memory and feeling; they sensitize systems that tell us when something is novel and when to remember it. (Nichols and Chemel 2006, 26)

The common subjective experiences elicited by these drug-related changes in brain systems include "altered perception of reality and self; intensification of mood; visual and auditory hallucinations, including vivid eidetic imagery and synesthesia; distorted sense of time and space; enhanced profundity and meaningfulness; and a ubiquitous sense of novelty" (Nichols and Chemel 2006, 3).

Whether these experiences are interpreted as a psychedelic "trip" or as a deep spiritual awakening may be influenced by one's experience-based expectations, the setting in which the drugs are taken, and the cognitive/ theological network out of which one provides a post hoc interpretation of the experience. These insights from neuropsychology and neurochemistry need to be considered when evaluating any claims of privileged religious knowledge based primarily on such experiences.

An extreme example of brain-related spiritual events is the so-called out-of-body experience—a brief episode in which a person's conscious self seems to leave the body and observe the body and the situation from a different spatial perspective. Within some Christian groups such experiences are looked upon as extremely valuable and desirable. For example, the members of one religious group in the United States (in Oregon) combine Christian and indigenous Brazilian religious beliefs. They recently won the right to import and brew a hallucinogenic tea for their religious services. The tea is brewed with the ayahuasca plant, which contains trace amounts of the chemical *dimethyltriptimine* (DMT), known to cause psychedelic phenomena including visual and audio sensations characteristic of an out-of-body experience (Siegel 2015). The resulting paradox: *even "out-of-body" experiences are "embodied" and "embrained" in the altered activity of brain networks.*

## Does Neurotheology Have Value?

As can be seen from these examples, research in neurotheology has produced some interesting results showing the tight link between spiritual experiences and the physical and physiological state of the brain. However, not everyone is impressed by these findings. Jeremy Groopman, a distinguished Jewish physician, wrote, "Why do we have this strange attempt, clothed in the rubric of 'neurotheology,' to objectify faith with the bells and whistles of technology? Man is a proper subject for study in the world of science. . . . God is not" (Groopman 2001, 165).

Similar views were echoed by Mario Beauregard, who works in the departments of radiology and psychology at the Université de Montréal. As reported by Christopher Stawski, Beauregard stated, "Obviously, the external reality of God can be neither confirmed nor disconfirmed by delineating neural correlates of religious/spiritual/mystical experiences. In other words, the neuroscientific study of what happens to the brain during these experiences does not tell us anything new about God" (Stawski 2004). Perhaps Beauregard is correct. Perhaps neuroscience cannot tell us anything about God. But in our view, it can tell us something about the way humans experience God and feel God's presence in their lives.

## EMBEDDED SPIRITUALITY

### Spirituality Is Embedded in the Context of Life

Except in very rare instances such as the lonely hermit, the spiritual dimensions to life and experience are lived out in community—fully embedded in our physical, cultural, and social environments. To understand the different spiritual and religious choices that each person makes, we must recognize that spirituality is embedded within a specific life context that includes the person's developmental history and friendship networks, as well as the broader cultural environment.

That raises a practical issue: How can churches best recognize and respond to the spiritual needs of persons who are physically embodied and socially embedded? In other words, what new insights can the advances in neuropsychology and social psychology provide about how churches should be organized and how worship should be conducted? Brown and Strawn argue that the embeddedness of spirituality requires that churches take seriously the need to combine faith with action:

> Remember, it is not just that what is thought or experienced occurs in the brain and is expressed in the body, but the impact goes the other direction as well: *actions influence thought.* . . . What we do with our bodies has a profound influence on what we think. . . . Thus, to participate in the Eucharist during worship (an extended form of gesture) is for this bodily activity to have a deep influence on our thoughts, feelings, beliefs, and future behavior quite beyond what is said. So, our argument for the role of participation and action in worship is an argument based on the profound embodiment of all thought. Without concurrent action, thought and belief is likely to degenerate into nothing but intellectualism, and worship into mere feelings. Formation of persons will be minimal. (Brown and Strawn 2012, 152)

This line of thought leads Brown and Strawn to propose that the commonly held assumption of body-soul dualism actually undermines the message of Jesus that calls Christians to action in the world for the sake of God's kingdom. They write:

> Our premise has been that viewing persons as bodies, not souls inhabiting bodies, *is truer to Scripture, as well as more resonant with modern neuroscience and psychology.* Most Christians believe that humans are souls that *have* bodies, not that we *are* bodies. They presume that the "real me" is not their body or even their behavior, but is something inside them, in their head or heart—in their mind or soul. Thus, it is

possible to be spiritual inside, without being religious in what we do—without participating in the communal religious life of the church. However, Christian life has a very different feel if the essence of the human person is not a ghostly, immaterial soul or spirit that is temporarily trapped in a fleshly body and hidden from view, but is the indivisible composite of the behavior, habits, thoughts, emotions, and personality of the physical body itself. (Brown and Strawn 2012, 159)

## Spirituality Is Both Embodied and Embedded

While we have treated our discussions of embodied and embedded spirituality in separate sections of this chapter, Warren Brown and Brad Strawn claimed that embodied and embedded spirituality are in practice intimately interrelated. In a review article describing recent advances in what is known as "extended cognition," they wrote:

> While mind is embodied more than just "embrained," it is also extended beyond the body in varying times and manners. . . . Thus, it is not simply that persons are embedded within cultural and social contexts, the contents and implications of which become internalized in ways that shape the nature of mental processes. Rather, the claim of situated and extended cognition is that important aspects of mental processing take place in reciprocally interactive exchanges with certain artifacts or persons within the current context, such that mind cannot be understood as simply internal. (Brown and Strawn 2017, 415)

Applying these views to the topic of spirituality, they write:

> Within Christian discussions about human nature, the biggest problem with crediting the claims of non-reductive physical-

ism and embodiment is not so much about the nature of mind as it is about the nature of spirituality. Does spirituality not require the existence of a nonmaterial, spiritual soul within each person? We have explored elsewhere the nature of Christian life when viewed from the point of view of the embodiment of personhood. We argued that "spirituality" should be reckoned as an awareness of the presence of God's spirit, rather than the cultivation of an inner non-material spirit or soul that is, by definition, private—separated from the external and social world. From the point of view of physicalism and embodiment, the world of spirituality moves outward by means of actions and interactions of the entire person with the world in a manner that encompasses the spirit of God. (Brown and Strawn 2017, 417)

### Spirituality Is Embedded in Communal Space

Because spirituality is embedded in the physical and social context of the person, "when persons are involved in congregational worship or corporate prayer, it becomes unclear who owns the spirituality in play" (Brown and Strawn 2017, 418). This raises some interesting questions about congregational life:

> What would be the implications for the life of the church if we took into account the possibility that spirituality can exist in extra-personal extended interactions? What if what we refer to as "spiritual" is not just embodied within the individuals? If spirituality is a property that emerges and exists in inter-individual space, at least during times of reciprocal and corporate activity such as worship, prayer, study, conversation, and work, then we may need to think differently about ecclesiology. (Brown and Strawn 2017, 418)

This account of "corporate spirituality" or extended spirituality permeating the shared worship space sounds a lot like the communal worship of the earliest Christians. However, contemporary Christians seem to be moving in the opposite direction, as Brown and Strawn note:

> However, as society and the church become ever more individualistic in their understanding of spirituality and faith (for example, the current movement toward ascetic spirituality), many parishioners have a difficult time giving an account of why the corporate activity of the church is important and what role it plays in Christian life. It may be seen as a "spiritual refueling stop," or as an institution for religious education, or simply as a source of accountability for one's individual spiritual quest. (Brown and Strawn 2017, 418)

There is much to digest in these provocative suggestions from two well-informed Christians at the cutting edge of both neuropsychology and practical theology. Their thinking is another illustration of how, by reflecting on the wider implications of developments in psychological science, we may discover new insights into how to live effectively as Christians in an ever-changing world.

## Impaired Spirituality

### Alzheimer's Disease

What happens to spirituality when the brain goes awry? Some of the most important and relevant insights from neuropsychology come from research on how Alzheimer's disease influences spirituality. Any belief that our spirituality is securely protected within a nonphysical part of us (the soul) is challenged by the experiences of individuals who have developed Alzheimer's disease. Some deeply religious people have suffered agonizing distress as they experienced the fragmentation and loss of precious aspects of their religious life. Such distress is equally painful for their loved ones and caregivers.

Glenn Weaver, who developed a large research program studying spirituality in Alzheimer's patients, has described some of the changes in these people's experiences of spirituality, religious faith, and meaning in life (Weaver 2004). The spiritual consequences of Alzheimer's dementia may vary widely across individuals. For some, the loss of independence and control leads to a greater dependence on God. For others, the gradual deterioration of cognitive abilities reduces spiritual interest. Weaver described in detail the experiences of Robert Davis, a Presbyterian minister who was diagnosed with Alzheimer's dementia when he was fifty-three and at the height of his ministerial career. With the help of his wife, he wrote a remarkable account of his experiences well into the middle stages of the disease. How his progressive brain disease affected his spirituality is graphically illustrated in his own words. He wrote:

> My spiritual life was miserable. I could not read the Bible. I could not pray as I wanted to because my emotions were dead and cut off. There was no feedback from God the Holy Spirit. My mind could not rest and grow calm but raced relentlessly, thinking dreadful thoughts of despair. . . . I can no longer be spiritually fed by sermons. I can get the first point of the sermon and then am lost. The rest of it sends my mind whirling in a jumble of twisted unconnected ideas. Coughing, headache and great discomfort have attended my attempts to be fed in all the ways I am accustomed to, meeting God through his Word. . . . My mind also raced about, grasping for the comfort of the Savior whom I knew and loved and the emotional peace that He could give me, but finding nothing. I concluded that the only reason for such darkness must be spiritual. Unnamed guilt filled me. Yet the only guilt I could put a name on was failure to read my Bible. But I could not read, and would God condemn me for this? I could only lie there and cry "Oh God, why? Why?" (Weaver 2004, 89)

This account should open our eyes to the important role of brain function for spirituality and religious behavior. It is difficult to see how a dualist perspective can accommodate the experiences of patients like Robert Davis. A "soul" or "mind" that is completely separate from the brain and not dependent upon neural functioning should not change (becoming more spiritual or less spiritual) as neurons die and brain tissue deteriorates.

## Parkinson's Disease and Religion

At Boston University School of Medicine, Patrick McNamara and his colleagues have conducted a series of pioneering studies that have provided new insights into religiosity in patients suffering from Parkinson's disease. With the increase in longevity in the general population producing a corresponding rise in the incidence of late-onset Parkinson's disease, it is important from a pastoral point of view to be aware of any changes in a person's religiosity and spirituality that may accompany the development of Parkinson's disease.

Parkinson's disease is a condition with a clearly defined pathology of the brain involving changes in the activity of dopaminergic neurons. Patrick McNamara, Raymon Durso, and Ariel Brown (2006) examined the role that prefrontal dopaminergic networks play in maintaining religious beliefs and behaviors. They found that, compared to age-matched controls, individuals with Parkinson's disease scored lower on measures of religious behaviors and expressed less interest in spiritual or philosophical issues. These findings raise the possibility that normal dopamine levels are important for maintaining religious motivation and goal-directed behavior based on religious beliefs and values. Follow-up studies revealed differences in religiosity between Parkinson's patients whose disease symptoms had appeared first on the left side of the body and those who showed earlier onset on the right side, suggesting that the two hemispheres of the brain play slightly different roles in religious behaviors (Butler, McNamara, and Durso 2011).

However, in a careful review of this area of research, Clare Redfern and Alasdair Coles (2015) have pointed out that interpreting the results of such studies is not a straightforward matter. Parkinson's disease produces a wide range of physical, cognitive, and emotional changes, and these symptoms vary in unpredictable ways as the disease progresses and as new treatments are introduced. This makes it difficult to claim that Parkinson's disease itself has a specific negative impact on spirituality.

Take a very simple example. With the progression of Parkinson's disease, mobility is severely limited. Therefore, it is not surprising if a person with Parkinson's shows reduced attendance at worship services. The effort of getting dressed and getting into the worship space may be overwhelming. Thus, a decline in religious behaviors by a Parkinson's patient would not necessarily imply any loss of spirituality or religious belief.

In addition, some research has found a more positive pattern of changes. At least some Parkinson's patients report a *strengthened* relationship with God and a more intentional search for religious meaning as a way of coping with their disease (Redfern and Coles 2015). These contradictory findings demonstrate the need for rigorous, well-controlled studies mobilizing all the combined skills of neurologists, psychologists, and theologians.

There is clearly much work to be done, but the existing research has already provided new insights into how spirituality may be affected in patients with Parkinson's disease. With a better understanding of what is happening in and to such persons, pastoral care may be improved.

### Peak Religious Experiences and Temporal Lobe Epilepsy

There is a significant literature in clinical neurology suggesting that some individuals with temporal lobe epileptic seizures experience intense religious states as a part of the "aura" preceding a seizure. These episodes of intense religious awe, ecstasy, or ominous presence appear to be a product of the abnormal electrical activity of the brain produced by their seizures. Although such cases are rare, they happen often enough to suggest something about the physical processes that may be associated with ordinary religious experiences.

There are a number of literary allusions to this phenomenon. Fyodor Dostoyevsky (who himself had a seizure disorder) gave a particularly graphic literary description of this sort of seizure in his account of the experiences of Prince Miskin in *The Idiot*. In this passage, Dostoyevsky describes (in the thoughts of Miskin) the sort of religious experiences that are sometimes associated with temporal lobe seizures:

> He fell to thinking that in his attacks of epilepsy there was a pause just before the fit itself . . . when it seemed his brain was on fire, and in an extraordinary surge all his vital forces would be intensified. The sense of life, the consciousness of self were multiplied tenfold in these moments. . . . His mind and heart were flooded with extraordinary light; all torment, all doubt, all anxieties were relieved at once, resolved in a kind of lofty calm, full of serene, harmonious joy and hope, full of understanding and the knowledge of the ultimate cause of things. . . . If in that second—that is, in the last lucid moment before the fit—he had time to say to himself clearly and consciously: "Yes, one might give one's whole life for this moment!" then that moment by itself would certainly be worth the whole of life. (Dostoyevsky 1969, 243)

A recent literary reference to this phenomenon can be found in Mark Salzman's modern novel *Lying Awake*. Salzman writes about a nun with religious visions associated with temporal lobe seizures (Salzman 2001). Accounts of religiouslike experiences associated with a temporal lobe seizure can also be found in the modern neurological literature. For example, this is the way one patient described the aura preceding a seizure: "Triple halos appeared around the sun. Suddenly the sunlight became intense. I experienced a revelation of God and of all creation glittering under the sun. The sun became bigger and engulfed me. My mind, my whole being was pervaded by a feeling of delight" (Naito and Matsui 1988, 123–24).

These clinical reports have led neuroscientists such as V. S. Ramachandran to speculate about the possible existence of a "God Module" in the brain—that is, "dedicated neural machinery in the temporal lobes concerned with religion" (Ramachandran and Blakeslee 1998, 187–88). However, Jeffrey Saver and John Rabin provide a more cautious view of the "temporolimbic hypothesis" of religious experiences, claiming that temporal lobe activity by itself is not enough to generate the sense of spiritual ecstasy. They hypothesize that increased activity within the limbic system (the major emotion center of the brain) tags certain encounters as "crucially important, harmonious, and/or joyous, prompting comprehension of these experiences within a religious framework" and providing "the sense of having touched the ultimate ground of reality and the sense of the unutterability of incommunicability of the experience" (Saver and Rabin 1997, 507). Whatever the most appropriate explanation of the meaning of this phenomenon, it is clear that *certain patterns of electrical activity involving the temporal lobes and the limbic system (sometimes occurring during a seizure) can cause intense, personally significant experiences that some persons describe as religious.*

## A Pause for Reflection

Psychiatrist Gaius Davies, after having discussed some of the effects of Alzheimer's disease on spirituality, continued:

> Many other illnesses where body and mind are affected together or reciprocally show quite vividly to many that the soul and spirit are embodied, and we cannot consider one aspect of the things of the spirit without bearing in mind always how much such things are affected by physical well-being. What used to be called "the cure of souls" was a phrase . . . which covered a wide spectrum of care—centuries

before any specific remedies were available such as antibiotics, anticonvulsants or antidepressants. The whole person had to be cared for by those called to the task. Now there are medical advances, and the technologies of surgery, anesthetics, and organ replacement, that have made a world of difference. . . . The care and cure of souls thus seeks to make a comeback as holistic medicine. At its best, such an approach has important components from medicine and from spiritual sources, as well as from complementary methods, often of ancient lineage but now offered in modern guise. (Davies 2004, 137)

It would be all too easy for Christians, after hearing about the latest discoveries by neuropsychologists, to wonder whether some of their most treasured beliefs and experiences have now been eroded and "explained away" or shown to be "nothing but" this or that aspect of human brain functioning. One natural response to such a situation would be to embrace a *warfare/conflict view* of the relations of science and Christian faith and to seek to defend traditional views of spirituality (and the existence of the "soul") against these latest attacks. Instead, we have suggested that a more constructive view sees the efforts of scientists as *affording new insights* into ancient truths that potentially have *the capacity to enrich* our understanding and experience of living as Christians in today's world.

But there is another danger we need to be aware of, namely, seeking to produce a concordism between the current science of spirituality and scriptural descriptions of the spiritual aspects of the Christian life. It would be unwise to believe that *today's science* and *today's biblical interpretations* are carved in stone. It would be even more unwise to produce a convincing concordism between the two that can then be appealed to, relying on the high status of science to bolster Christian beliefs. To take that path would be to misunderstand both the nature of the scientific enterprise and the nature of developing biblical scholarship.

# CHAPTER 8

# Insights about Conversion, Morality, Wisdom, and Memory

R ELIGIOUS CONVERSION has, for more than a century, been among the topics most frequently discussed and energetically studied by psychologists interested in the psychology of religion. With the development of improved noninvasive techniques for studying brain activity, religious conversion has also attracted the attention of neuroscientists. In this chapter, we discuss religious conversion as another area in which scientific research has provided fresh insights that have the potential to enrich our understanding of an important aspect of Christian faith and life. But remember that the enrichments can flow in both directions. Recent biblical scholarship has also provided new insights that could influence how psychologists approach the topic of conversion.

These new methods for studying the brain have also shed new light on moral decision-making, especially the changes in moral reasoning that occur after selective brain damage. We will point out the relevance of psychology and neuroscience in affording a deeper, more sympathetic understanding of what might be happening when individuals suddenly act "out of character," violating the expectations of Christian living.

Moral reasoning is closely related to another aspect of cognition that scripture describes as wisdom. We highlight here some insights from psychological research on wisdom that have a bearing on living a Christian life in the modern world. Finally, wisdom is also related to remembering the past and learning from it. We conclude this chapter with a discussion of the extensive research on human memory. Because the Bible contains many episodes that involve the testimony of eyewitnesses, we believe that the evidence from psychological science is directly relevant to biblical scholarship and faith affirmations about the reliability of scripture.

## UNDERSTANDING RELIGIOUS CONVERSION

### Resistance to Studying Conversion

When those outside the Christian community look at the phenomenon of religious conversion—either in terms of a nonbeliever coming to faith, or a nominal believer becoming a deeply committed believer—they often dismiss the event with statements such as, "Conversion is just emotionalism," or "Conversion is all psychological." Understandably, this reaction upsets Christians who attach great importance to the experience of conversion as the starting point, or perhaps the turning point, in a person's religious journey. Any suggestion that our understanding of any religious behavior, including conversion, might be enhanced by psychological research evokes a negative response. "Such experiences are too sacred to be investigated by psychologists! The very act of studying the religious event would undermine or destroy the experience." While one may sympathize with this view, we do not agree with it.

One reason for apprehension about scientific research on religious experience and behavior is the assumption that to produce a psychological explanation of some aspect of religious behavior is to "explain it away." By implication, this suggests that a psychological explanation removes any necessity to take seriously the religious beliefs of the person being studied. The fallacy behind such an attitude is illuminated by a story reported by Gordon Allport: A student, after hearing Archbishop William Temple give an address at Oxford, opened the discussion by commenting, "Well, of course, Archbishop, the point is that you believe what you believe because of the way you were brought up" (knowing that he was the son of an Anglican clergyman). The archbishop quietly and coolly replied, "That is as it may be. But the fact remains that you believe that I believe what I believe because of the way I was brought up, because of the way you were brought up" (Allport 1950, 124). Rather than accepting the student's criticism as "explaining away" his beliefs, the archbishop pointed out that the student's comment is the entry to an infinite logical regression—intended to avoid taking seriously the views of the other person.

## The Importance of Studying Conversion

One important reason for taking conversion seriously is that, according to the founder of the Christian faith, conversion is not an optional "extra." Rather, it is part of the essence of being a Christian. Jesus said, "Unless you are converted . . . you will not enter the kingdom of heaven" (Matthew 18:3 MEV). Other translations of this passage use "turn" or "change" in the place of "convert," to emphasize the need to move in a new direction. As we shall see later, New Testament scholar Joel Green demonstrates convincingly that, according to scripture, conversion is not just a one-time event, but a lifelong journey that completely transforms the person (Green 2015).

A review of the literature on the psychology of religion shows that conversion is a common topic, but the depth of coverage varies widely. In some books and articles, such as Koteskey (1980), Myers and Jeeves ([1987] 2003), Jeeves (1988), or Moes and Tellinghuisen (2014), religious conversion is mentioned only briefly in discussing other topics. Some books devote a chapter to conversion, as in *Psychology and Christianity: The View Both Ways* (Jeeves 1976) and *The Psychology of Religion and Spirituality* (Sizemore 2016). In other books, such as *The Oxford Handbook of Religious Conversion* (Rambo and Farhadian 2014), the topic of conversion takes center stage. In *Battle for the Mind: A Physiology of Conversion and Brain-Washing*, psychiatrist William Sargant (1957) focused on sudden conversions in a range of North American religious sects, such as the snake handling sects. He speculated about some of the possible brain mechanisms involved when dramatic conversions occurred, basing his theorizing on the work of the Russian physiologist Ivan Pavlov. David Wells' book *Turning to God: Biblical Conversion in the Modern World* (Wells 1989) leaves no doubt about the centrality of this topic for Christians and believers in other religious faiths, as well as the relevance of studying the psychological mechanisms that are at work during conversion experiences.

We should point out that the early research on the psychology of religion—and particularly the research on conversion—occurred before or in the initial stages of the so-called "cognitive revolution" which transformed psychology. Linked with this cognitive revolution were rapid advances in understanding the relationship between psychological processes and brain mechanisms (which were reviewed in chapter 6). Also, because conversion often occurs within a social group, an understanding of the psychological processes of social behavior, along with recent work in the field of "social neuroscience," is clearly relevant. For example, the chapter headings in the book *Social Neuroscience* (Cacioppo and Bernston 2005), such as "The Brain Determines Social Behavior," "Neural Mechanisms of Empathy in Humans," and "Exploring the Neurological Substrate of Emotional and Social Intelligence," indicate potential connections to the experience of religious conversion in a social environment.

Thus, it is important to look again at conversion as an example of an aspect of religious behavior and spiritual life for which new insights from cognitive science, neuropsychology, and social neuroscience are now available, providing enrichments to our understanding of the life journey of a believer.

*Insights about Conversion from Modern Biblical Scholarship*
In this book, we try to give attention both to advances in psychological science and advances in biblical scholarship. We are fortunate that Joel Green, a leading New Testament scholar who is also remarkably well-informed and familiar with the literature in psychology and neuroscience, has shared with us his views on conversion in the Gospel of

Luke and in the book of the Acts of the Apostles (Green 2015). Green recognized that, while we should value the groundbreaking work by William James in his *Varieties of Religious Experience* (James 1902), we should nevertheless recognize that an uncritical acceptance of his views may actually prevent us from benefiting from the new insights coming from both ongoing scientific research and dedicated biblical scholarship. After referring to developments in cognitive science, neuropsychology, and neuroscience, Green wrote:

> James articulates his view that religious experience is primary for human nature and gives rise to myriad theologies, philosophies, and religious institutions. Primary, then, are *"the feelings, acts and experiences of individual men in their solitude, so far as they pretend themselves to stand in relation to whatever they may consider the divine."* Given this dual emphasis on individuality and interiority, James's definition of "conversion" is as unsurprising as it may be familiar, at least in its broad strokes. (Green 2015, 6)

In James' view, what is the process of conversion? He wrote:

> To be converted, to be regenerated, to receive grace, to experience religion, to gain an assurance, are so many phrases which denote the process, gradual or sudden, by which a self hitherto divided, and consciously wrong, inferior and unhappy, becomes unified and consciously right, superior and happy, in consequence of its firmer hold upon religious realities. This at least is what conversion signifies in general terms, whether or not we believe that a direct divine operation is needed to bring such a moral change about. (James 1902, 189)

Green noted that James had a view of conversion that was rooted in a Western individualism. He wrote, "James articulates well what is axiomatic for many, namely, this stress on individual oriented, feeling based, interior religion" (Green 2015, 8). Green then invited us, taking note of a century of advances in psychology and biblical scholarship, to be ready to question James's understanding of conversion even though "it seems right at an intuitive level" (Green 2015, 8). Green urged us to reflect that

> James's psychological description of conversion finds a ready home in this understanding of the human person, so it is critical to recognize that this "modern self" is not only modern but non-Eastern—or, to put it more succinctly, this description of the "self" is neither trans-historical nor transcultural. Nor is it particularly biblical. (Green 2015, 9)

According to Robert Di Vito, this modern self contrasts sharply with the Old Testament view of a human who

> (1) is deeply embedded, or engaged in social identity; (2) is comparatively decentered and undefined with respect to personal boundaries; (3) is relatively transparent, socialized, and embedded (in other words, is altogether lacking in a sense of "inner depths"); and (4) is "authentic" precisely in its heteronomy, in its obedience to another and dependence upon another. (Di Vito 1999, 221)

Klaus Berger painted a similar portrait of New Testament anthropology, in which a human's personal identity or self is "outer-directed," embedded in the social interactions within one's community (Berger 2003). Reflecting on these recent contributions, Green concluded that "with such emphases as these, we have moved a country mile away from the mainstay of James's definition of conversion: a 'self' experiencing subjective crisis and inner resolution" (Green 2015, 10).

## INSIGHTS ABOUT CONVERSION FROM RECENT SCIENTIFIC INVESTIGATIONS

Given the vast differences between the modern world and the world of the Bible, do the methods and assumptions of the modern social sciences have any relevance for biblical scholarship? Joel Green recognized the importance of sociology when he wrote:

> I take it as axiomatic that sociological work has much to offer our understanding of conversion in the NT materials, and that the work of Peter Berger (which synthesizes earlier work in the sociology of religion) in particular has much to offer, with its emphasis on the religious ordering and reordering of reality. (Green 2015, 11)

Green, however, was also aware of the limitations of sociological analysis: "However fully human beings can be understood in sociological terms, there is always more to the human story than sociology can recount" (Green 2015, 11). Green continued:

> Nor can psychology be so easily set aside, even if one might wish to set aside a particular psychological approach to conversion articulated by William James. . . . Psychology will not go away. Human beings and their experiences in the world, even their religious experiences in the world, are not part cerebral and part social but fully integrated in their embodiment. (Green 2015, 12)

Green next takes up the question of how cognitive science can illuminate the description of conversion in the narrative of Luke-Acts. In defending his reliance on cognitive science, Green writes:

The essential characteristic of a cognitive approach is its non-negotiable emphasis on embodiment—that is, "the role of an agent's own body in its everyday, situated cognition," its irreducible emphasis on somatic existence as the basis and means of human existence, including religious experience and the exercise of the mind. . . . The capacities that allow humans to experience God are themselves embodied and subject to study by means of the cognitive sciences. (Green 2015, 19)

However, Green noted the importance of proceeding cautiously:

Our agenda, then, is not to transform the narrator of Luke–Acts into a cognitive scientist. Nor do we assume that Luke was an "anonymous cognitive scientist," as though he were working with cognitive categories unknowingly. Instead, a cognitive approach functions as a place to stand in order to survey the landscape of the Lukan narrative, a viewing point that allows vistas unavailable to those whose starting point has been determined by William James. (Green 2015, 21)

Green described a series of studies in contemporary psychology, sociology, and cognitive science that allow a different, non-Jamesian view of conversion to emerge—one that is more faithful to the conversion narratives in the Gospel of Luke and the Acts of the Apostles (Green 2015, 22–37). As examples, Green referred to

- ► Work by Antonio Damasio in interpreting the effects of selective brain damage, beginning with the description of the well-documented case of Phineas Gage, whose personality changed after damage to the frontal lobes.
- ► Neuroscience research on the role played by environmental factors in brain development during infancy and childhood.
- ► Studies by Eleanor Maguire and her colleagues on how the brains

of London taxi drivers selectively change as a result of memorizing the location of streets and buildings throughout London.

► Research on the Capgras syndrome (in which familiar people are judged as "impostors") and "alien hand syndrome" (in which people fail to recognize a body part as belonging to them).

► Studies by Andrew Newberg on blood flow to specific areas of the brain during glossolalia, as well as his research on other neural correlates of spiritual experiences.

After assembling support for the relevance of developments in psychology and neuroscience for our understanding of conversion, Green concluded:

Converts are those who, enabled by God, have undergone a redirectional shift and now persist along the Way with the community of those faithfully serving God's eschatological purpose as this is evident in the life, death, and exaltation of the Lord Jesus Christ, and whose lives are continually being formed through the Spirit at work in and through practices constitutive of this community. (Green 2015, 164)

According to Green, the evidence of scripture—together with fresh understandings from psychology, sociology, and cognitive science— suggests that true conversion is manifest *not in a single event* but in a life walking what the New Testament narratives describe as "The Way" in community with fellow Christians—a life that exalts the Lord Jesus Christ and that is characterized by participation in an active, Spirit-filled community.

## UNDERSTANDING THE NEUROSCIENCE OF MORAL BEHAVIOR

It is widely presumed that moral behavior stems from one's worldview and religious perspectives; thus, religion and morality go hand in hand.

If you hold a body-soul dualism view of the person, processes of moral decision-making should happen primarily in the domain of the soul, not the body. In contrast, a monist view of the person would presume that it is the brain/body (or better, the whole "person") doing the deciding. What light can psychological science throw on this issue? What does recent research tell us about the brain systems and processes that contribute to the moral regulation of behavior?

### The Frontal Lobes and Moral Behavior

Arguably the most famous single case in all of neurology is Phineas Gage. The brain injury that Gage suffered did not kill him, but it did change his behavior, including his moral decision-making and impulse control. The outcome of his case illustrates the important role of the frontal lobe of the brain, particularly the lower-middle portion of the frontal lobe known as the *orbitofrontal cortex*, in moral behavior.

Gage received major damage to his frontal lobes when an iron bar that he was using to tamp an explosive charge blasted through his eye socket and out of the top of his head. While Gage never lost consciousness, and seemed to have recovered physically within days, he was never the same person. Prior to the accident, he was an intelligent person, a capable and efficient worker, an excellent manager, a responsible family man, and an upstanding citizen. While he maintained his general intelligence after the accident, damage to his frontal cortex resulted in an interpersonal style best described as unreliable and capricious, socially inappropriate, and amoral (Larner and Leach 2002; Macmillan and Lena 2010).

Studies of other patients with this form of brain damage show that they typically have difficulty regulating their behavior in order to abide by norms of socially acceptable or moral behavior. Such individuals may, capriciously and without malicious intent, violate social conventions, laws, ethical standards, or the rules of courtesy, civility, and regard for the benefit of others.

Antonio Damasio has carried out extensive studies of individuals with injury to the orbitofrontal cortex and has proposed his "theory of somatic markers" as an explanation of what goes wrong in this sort of brain injury (Damasio 1994). According to this theory, our experiences with the contingencies of life cause us to develop anticipatory emotional responses in certain situations. The responses are largely unconscious, but they produce physiological changes, such as increased heart rate, muscle tension, or sweaty palms. These bodily emotional responses emerge into consciousness in the form of subtle feelings and intuitions, such as a feeling of suspicion toward a person, or the feeling that a planned action might not be the right thing to do. These anticipatory responses guide our behavior, particularly when we lack conscious knowledge and experience that is relevant to the situation.

Damasio believes that damage to the orbitofrontal cortex decouples conscious processes from this autonomic response system. In these patients, the process of mentally considering an action that they know (unconsciously, based on prior experiences) is wrong does not evoke negative autonomic responses. Without these evaluative emotional responses, behavior loses its anchor in previous experiences and becomes capricious. The implication of Damasio's theory is that part of what we "know" about the world is only available to our conscious decision-making by way of our emotions.

*Dynamic Brain Activity during Moral Decision-Making*
A rapidly developing subfield within neuroscience involves mapping the brain areas involved in different forms of interpersonal, economic, and moral decision-making. The general approach in this form of research is to have persons engage in decision-making tasks while their brains are being scanned using functional magnetic resonance imaging (fMRI). For example, a number of studies have demonstrated the activation of the limbic (emotional) areas of the brain during what one would presume to be tasks requiring merely the "cold" calculation of the likelihood of financial gains and losses (Basten et al. 2010). Limbic involvement is particularly intense when the financial decision also involves interpersonal variables such as trust.

Joshua Greene and his collaborators have studied moral decision-making using similar techniques (Greene et al. 2001). First, they merely observed the enhancement of activity in different brain areas as the moral dilemmas that were presented became progressively more difficult. They found that the lateral frontal lobes and limbic areas became more active as moral decisions became more difficult. A follow-up study involved moral dilemmas that required one to imagine either directly inflicting harm on one person in order to save the lives of many other persons, or indirectly allowing harm to come to one person in order to save the others (Greene et al. 2004). Functional brain imaging indicated that having to choose to directly harm another person in order to save many others was correlated with activation of a different network of brain areas (including the medial frontal cortex and parietal lobes) from those activated by imagining a decision involving indirectly allowing harm. The additional brain areas activated by decisions about directly harming someone are areas typically involved in modulation of social action and representations of the self.

The general finding from this kind of research is that moral regulation of behavior is an *embodied* process, and that different forms of moral decision-making involve different patterns of brain activity. The somatic marker theory (Damasio 1994) suggests that one important contribution to moral behavior comes from the feelings elicited during interpersonal encounters—both feelings toward others (e.g., empathy, compassion, or trust) or feelings about the interpersonal nature of a situation (e.g., unfairness or social isolation).

### Brain Activity during Social Interactions

Since trust and interrelatedness are salient features of the religious life, it is noteworthy that these also are firmly embodied in the activity of brain networks. For example, Capgras syndrome is a disorder of the experience of familiarity when encountering close friends and family (Sinkman 2008). In these rare cases, damage to parts of the temporal lobe produces a disorder characterized by the individual's conviction that close friends or family members are actually "doubles" or impostors. For example, someone suffering from Capgras syndrome may visually recognize a family member but say, "That person looks like my wife." Because of the damage to the temporal lobes, the person with Capgras syndrome does not experience the accompanying feelings of familiarity and deep personal regard typically associated with the visual perception of a loved one. Therefore, the patient presumes that the family member or friend in question must be an impostor. Capgras syndrome demonstrates that feelings of familiarity and love can be dissociated from the visual recognition of the identity of the person by a dysfunction of the brain's temporal lobes.

Another important aspect of social interaction is the experience of being socially included or excluded from a group. Naomi Eisenberger and her colleagues have done functional brain scanning as individuals are being either included in, or excluded from, a game being played in a virtual reality group setting. The fMRI images show that the experience of social exclusion is accompanied by greater activation in the cingulate cortex (typically associated with negative emotions and anxiety) and the right frontal cortex (involving future planning and evaluation of cost/benefit ratio) (Eisenberger, Lieberman, and Williams 2003).

## Understanding the Neuroscience of Wisdom

### Biblical Views of Wisdom

In the Bible, morality and wisdom have a strong connection. Ecclesiastes 10:2 (GNT) claims, "It is natural for the wise to do the right thing and for fools to do the wrong thing," while James 3:17 (NRSV) reports, "But the wisdom from above is first pure, then peaceable, gentle, willing to yield, full of mercy and good fruits, without a trace of partiality or hypocrisy."

The Hebrew scriptural canon contains several books that belong to the genre called "wisdom literature"—Job, Proverbs, and Ecclesiastes, along with the Song of Songs (Song of Solomon) and a few of the psalms (Penchansky 2012). Song of Songs contains poems celebrating romantic love. Proverbs is a collection of wise sayings about living life in community, and several psalms have a similar style. According to Old Testament scholar Ronald Clements (Clements 2000), the other two books have highly distinctive features. Job is an extensive poem dealing with the problems of theodicy—justifying the ways of God to human beings and how to face the reality of undeserved human suffering. Ecclesiastes is unusual because it is the work of a single author giving his personal perspective, even though it is heavily steeped in the ancient language and traditions of wisdom.

As Christians, we all aim at manifesting wisdom—which Clements claimed is an approach to life rather than a set of pithy proverbs. He wrote, "Overall, wisdom sought rules that were directed towards shaping character and attitude, rather than offering anything akin to a law code. The goal is to create a wise person rather than to define the rules of what wisdom dictates should be done on particular occasions" (Clements 2000, 27).

### Wisdom Requires an Intact Brain

What possible insights into the development of wisdom, as described in scripture, could come from research in psychology? Neuropsychologist Warren Brown had some provocative thoughts on this topic, which he defined as "that aspect of intelligent activity that allows for successful problem-solving in widely different domains, but is particularly evidenced in negotiating psychosocial exigencies of everyday life" (W. Brown 2000, 195). Evolutionary biologist Jeffrey Schloss offered a similar definition that emphasized the connection to the realities of life:

> Wisdom is living in a way that corresponds to how things are. It is not mere knowledge, nor is it mere moral admonition, but it involves deep insight into the functioning, meaning, and purposes of existence, along with the ability to discern how to live correspondingly, that is, in accordance with the way things are. (Schloss 2000, 157)

Brown notes that most discussions of wisdom assume that persons being discussed have typical neuropsychological functioning. But what if the brain is not intact? There is considerable evidence that brain dysfunction reduces a person's capacity for demonstrating wisdom.

Brown pursued this question in depth, looking first at people who were born without the corpus callosum, a thick band of neurons that passes information between the left and right hemispheres of the brain. Persons who are missing this structure often have other brain abnormalities that lead to moderate to severe cognitive disabilities (Lassonde and Jeeves 1994). However, the absence of the corpus callosum can also occur without other brain abnormalities—in which case the individual's intelligence may be within the normal range (Brown and Paul 2000). But the inability to share information between the hemispheres leads to subtle deficits in complex problem-solving, including the interpersonal problems of daily life. Brown concluded that, in patients without a corpus callosum, "failure to adequately understand complex social dynamics leads to poor social judgment, which can be understood as diminished or absent wisdom" (W. Brown 2000, 203). Ironically, this deficit includes impaired understanding of proverbs about wisdom.

Brown also examined research on patients with damage to the brain's orbitofrontal cortex, including the case of Phineas Gage and the individuals studied by Antonio Damasio. These patients tend to show poor management of their time and their money, flawed decision-making, and erratic, impulsive, socially inappropriate behavior. Interestingly, some of these patients performed well in laboratory tests of moral judgment and social problem-solving, but failed dismally in applying those principles to situations in their daily lives. As Brown wrote, "Wisdom is more than what a person says; it must also involve the ability to regulate behavioral choices in everyday situations" (Brown 2000, 209). This evokes the message of James 1:22 (NRSV): "But be doers of the word, and not merely hearers who deceive themselves." Based on this neuroscientific evidence, Brown concluded:

Wisdom is a concerted function of the entire brain (and body). It involves judging truly what is right or fitting and being disposed to act accordingly. The impact of specific forms of brain damage or abnormal brain development on judging and acting helps to enlighten us as to the various neural systems and cognitive abilities that contribute to the wisdom of persons. (W. Brown 2000, 210–11)

## Wisdom Develops across the Life Span

In addition to neuropsychology, other areas of specialization in psychology, such as developmental psychology (specifically the study of adolescence and adulthood), can also contribute to our understanding of the biblical concept of wisdom and of the factors that enable wisdom to be expressed in the lives of Christians at different points in their life cycle. Such insights have the potential to be helpful to anyone ministering to adolescents and young people in the local church. For example, James Furrow and Linda Wagener drew upon well-grounded research in psychology to examine the role of religion in the development of wisdom in adolescence (Furrow and Wagener 2000). Thus, once again, new insights into this pervasive biblical concept are made available through careful research by psychological scientists.

## MEMORY RESEARCH AND THE RELIABILITY OF SCRIPTURE

### The Biblical Narrative Contains Eyewitness Accounts

The reliability of scripture is a cornerstone of Christian doctrine. The main Christian denominations include in their affirmation of beliefs a statement about the authority of the Bible. On what do we base our trust in the reliability of the words of scripture? Some of the evidence about the origins of Christianity, as contained in the Gospels and Epistles of the New Testament, comes from eyewitness accounts of events in the lives of Jesus and his apostles. Thus, honest Christian students who are familiar with the psychological research on memory are faced with a dilemma: their faith is based on events recalled from human memory, yet there is considerable research demonstrating that memory is unreliable (e.g., Loftus and Palmer 1974). It remains a puzzle why, in books and articles relating psychology and Christian faith, one finds scarcely any attempt to help students think their way through this dilemma. Fortunately, biblical theologians have addressed this topic, demonstrating again that biblical scholarship and psychological science can provide mutual insights.

Richard Bauckham's widely acclaimed book *Jesus and the Eyewitnesses* dug deeply into the vast psychological literature on memory, carefully summarizing the attempts, past and present, to understand the processes of memory (Bauckham 2006). Bauckham drew heavily upon pioneering work by F. C. Bartlett on memory schemas (Bartlett 1932) and Alan Baddeley on working memory (Baddeley 1992), as well as research by Jerome Bruner and Carol Feldman on group narrative (Bruner and Feldman 1996). Bauckham's description of this psychological literature provides rich insights for anyone wishing to understand the part played by different types of memory in the reporting of eyewitness accounts and as used in producing the Gospel records. Interestingly, we find no hint of conflict or concordism in Bauckham's discussions.

### Can We Trust the Memory of the Biblical Eyewitnesses?

Bauckham used psychological research on memory to support the idea that the New Testament documents can be viewed as historical records supported by contemporary eyewitness accounts. His book has been criticized by biblical scholars who believe that the New Testament documents provide little historical evidence, because they were written and edited within religious communities more interested in promoting their theological perspective than in transmitting the "facts" about the life of Jesus (see Redman 2010 for a summary of the criticisms).

In a recent chapter, Bauckham acknowledged that the trustworthiness and accuracy of the biblical record is an area of very lively debate among biblical scholars and theologians. He writes:

> Some historical Jesus scholars have concluded, from an acquaintance with the literature on memory research, that memory is generally rather unreliable, grossly distorted by present concerns, and unstable. However, while this impression may be given by some of the research, the reason is that the psychologists often focus on memory's failings. They do so because understanding when and how memory sometimes fails may assist understanding of how it manages to be usually very reliable. (Bauckham 2015, 65)

Responding to his critics, Bauckham provided a review of the relevant research on memory, highlighting findings that address the reliability of biblical accounts of historical events. He reminds us that cognitive psychologists have studied several distinct types of memory, including short-term memory, working memory, and long-term memory, as well as procedural memory, semantic memory, and episodic memory, and notes that different views about the reliability of biblical eyewitness accounts sometimes arise because of a failure to distinguish among the different types of memory.

Research on memory is continuing to make progress on issues that are relevant to the biblical record. For example, a recent study challenged the perception that eyewitness memory is inherently fallible, finding that eyewitness confidence can reliably indicate the accuracy of an identification in some cases (Wixted and Wells 2017, 55). This finding is important in view of the extremely high level of confidence that early Christians showed in their memory of the events in Jesus's life, sometimes choosing a martyr's death rather than renounce what they believed to be the truth. The take-home message: keep up-to-date with psychological science, because it can provide insights into important topics in Christian faith and life.

## IMPLICATIONS AND INSIGHTS

### *Mental Events Are Embodied in Brain Activity*

The research we have just surveyed suggests that a better understanding of typical brain function, as well as the impact of brain damage, can illuminate religious experiences and behaviors, including religious conversion, moral reasoning, and spiritual wisdom. In other words, these spiritually focused mental events and cognitive processes are embodied in the activity of neurons within the lump of tissue we call the brain.

Such a view challenges the beliefs of many religious people who hold that spiritual experiences and moral decisions are manifestations of nonmaterial human minds or souls. Even though these people might rationally agree that our brains and bodies are somehow involved in those experiences, they feel strongly that religious experiences are not physical, and thus could not be the products of our physical bodies, or be influenced by the ordinary physiological processes of the brain.

This brings us back to one of the central issues in biblical anthropology: *dualism versus monism.* Wolfhart Pannenberg (1944, 182) framed the question this way: "When the life of the soul is conditioned in every detail by bodily organs and processes, how can it be detached from the body and survive it?" Pannenberg believed that the consequences of neurological damage and disease, as well as the evidence that specific brain functions are associated with every aspect of human behavior (including our moral decisions and spiritual experiences), point to the necessity of abandoning dualism. We believe that the subsequent decades of mind-brain research have further reinforced Pannenberg's conclusion.

How can moral responsibility and the value of religious experiences be preserved in the face of the scientific research we have just reviewed? The answer, we believe, is to return to a Hebrew-Christian view of the person advocated increasingly over the past century by biblical scholars—*a view that sees religious experiences and moral thoughts as embodied in physical brain activity and communicated to the world through the actions of a physical body* (Brown, Murphy, and Malony 1998; Green 2008).

*Mental Events Are Embedded in the Social Context*

We have argued that spirituality, along with all other mental activity, is firmly embodied in the physical reality of our cells and organ systems, and especially in the activity of the nervous system. In addition, we believe that religious experiences and behaviors are firmly embedded in a particular context, influenced both by the individual's internal personal beliefs at that time and by the external physical and social environment.

What evidence has emerged about the social embeddedness of spirituality? David Myers highlighted some of the evidence in a review of the literature on the link between social support and health (Myers 1998). This research suggests that environments—such as religious communities—that support our need to belong also foster stronger immune functioning. Social support improves health by reducing the level of stress hormones, thus boosting resistance to cold viruses, as well as calming the cardiovascular system and lowering blood pressure.

Myers pointed out that throughout history the two different healing traditions—religion and medicine—have joined hands in caring for suffering humans. At times those hands belonged to the same person. Maimonides, for example, was a twelfth-century rabbi *and* a renowned physician (Myers 1998). However, as medical science developed, medical healing and religious healing tended to diverge. No longer did people ask God to spare their children from smallpox, but instead took them to be vaccinated. Rather than seeking a spiritual healer in the face of bacterial fever, they administered antibiotics.

Surprisingly, this long-term trend of separating medicine from religion has been reversed somewhat in recent decades. In his review, Myers noted that more than one thousand studies have tried to correlate religious faith and religious practices with general health and the healing of specific illnesses. One of the most important of the studies compared the death rates of thirty-nine hundred Israelis living in two types of kibbutzim. Some lived in eleven religiously orthodox communities, and others in eleven matched nonreligious communities (Kark et al. 1996).

The results indicated that belonging to a religious community was associated with a strong protective effect against poor health.

How should we interpret these results, especially in light of the fact that members of both types of collectives enjoyed a high level of social support? Perhaps the religious beliefs of the religious kibbutzim members boosted the beneficial effects of social support for coping with illness and disease above and beyond that experienced within a nonreligious kibbutz. For the members of the religious collectives, there also appeared to be some protective effect of the enhanced well-being associated with a coherent worldview, a sense of hope for the long-term future, feelings of ultimate acceptance from God, and the relaxed meditation of prayer or Sabbath observance. Each of these may be seen as part of the spiritual dimension at work in healing, embedded in the social environment of the community of believers.

### Mutual Insights

We believe that the examples given in this chapter, along with others noted in Jeeves (2004; 2011; 2015b), further illustrate the benefits to be gained when psychologists, biblical scholars, and theologians come together to share their respective insights into what we know about human nature. There are new insights into Christian conversion, further emphasizing that it is *whole persons* who are being transformed by the call of God, not just one part of them that we could label the "soul." Advances in neuropsychology should help all of us feel greater sympathy for individuals whose moral reasoning has been diminished by brain dysfunction. New developments in neuroscience illuminate the biblical concept of wisdom and its role in our daily lives. Finally, there is new information from memory research to help biblical scholars consider the reliability of the eyewitness accounts of events documented in the scriptures.

# CHAPTER 9

# Insights from Evolutionary Psychology

## THE BACKGROUND

THE QUESTION could be asked, why devote a chapter to insights from evolutionary psychology when examination of the index to *Psychology and Christianity: Five Views* (E. Johnson 2010) shows no reference at all to evolutionary psychology? The answer is that the *Five Views* book and similar volumes mentioned in earlier chapters have tended to be very narrowly focused, dealing almost exclusively with a restricted set of areas, such as clinical psychology, counseling psychology, and pastoral psychology. In so doing, we have argued, they fail to provide the help needed by serious Christian students being taught evolutionary psychology and neuropsychology in university courses in psychology.

In the years since Charles Darwin wrote *The Descent of Man* (Darwin 1871), the relevance of evolutionary biology to psychology has been clearly demonstrated. Unfortunately, in spite of the fact that psychology had taken root more strongly in the United States than anywhere else in the world, this relationship was muddied by prolonged battles about the implications of evolutionary theory for what were claimed to be fundamental Christian beliefs about the nature and origin of humans. This, of course, is a topic that still features prominently in the headlines today, as illustrated by the attention being given to a book by Dennis Venema and Scot McKnight (2017) titled *Adam and the Genome*. In commenting on the publication of this book, Denis Alexander, former director of the Faraday Centre for Science and Religion at Cambridge University, highlighted the importance of cultural context when he wrote:

> The book also acts as a reminder to non-American readers, should one be needed, of the very different cultural and theological context in many Christian communities in other parts of the world. It is not that creationism does not exist outside North America—for it surely does—but rather the topic does not seem quite so fraught nor, in many cases, even a topic that attracts much attention. I was brought up in an evangelical home in Britain, but I cannot remember a time, at least since studying biology in high school, that I did not believe in human evolution. (Alexander 2017)

What is also surprising for some people is that some of the pioneers in the development of evolutionary psychology were themselves committed and practicing Christians. For example, William Thorpe, professor of zoology at Cambridge University in the second half of the twentieth century, published a book titled *Learning and Instinct in Animals* (Thorpe 1956), which was in effect a book on evolutionary psychology. Thorpe, himself a practicing Christian of Quaker persuasion, later gave the Gifford lectures at the University of St. Andrews under the title *Animal Nature and Human Nature* (Thorpe 1974). Nowhere in his writings is there any suggestion that he believed that there was warfare or conflict between evolutionary psychology and Christian faith, nor any attempts at constructing false concordisms.

## UNDERSTANDING EVOLUTIONARY PSYCHOLOGY

### What Is Evolutionary Psychology?

Evolutionary psychology refers to the study of the evolution of behavior and the mind using principles of natural selection. Our ancestors lived in a harsh environment. Genes that improved the chances of survival and reproduction were more likely to be passed on to future generations, while genes that interfered with survival and reproduction gradually disappeared from the human gene pool. Thus, natural selection influenced the development of the physical characteristics, behavioral tendencies, and information-processing systems that we see in humans today. We are the descendants of people who successfully solved the problems of survival and reproduction. The losers in the "game of life" left no legacy.

What is today called *evolutionary psychology* used to be called *comparative psychology*. It has an illustrious history in psychology's development. The focus of much earlier research in comparative psychology was the search for a way to link changes in sensory processes and capacity for learning to an animal's position on the phylogenetic scale. For example, in two earlier papers, one of us (MAJ) suggested a new metric for making meaningful cross-species comparisons (Rajalakshmi and Jeeves 1965; Rumbaugh and Jeeves 1966). This in turn led to attempts to link the increasing complexity of the brain and central nervous system with more elaborate behaviors and learning capacities. However, the findings of ethologists such as Konrad Lorenz and Nikolaas Tinbergen (see Burkhardt 2005) alerted us to the fallacy of believing that only with the increased complexity of the nervous system could there be an increase in learning capacities and the complexity of social behaviors. Some of the animals with the simplest nervous systems—such as ants and bees— showed remarkably complex forms of social behavior. These finding are completely uncontroversial within the scientific community, but *some within the general population fear that these discoveries undermine traditional theological accounts of what makes humans special, including the idea that humans are made "in the image of God."*

## Does Comparing Humans with Animals Demean the Nature of Humanity?

An endorsement of the book *Good Natured* (1996) by Frans de Waal, a leader in the field of evolutionary psychology, began:

From the beginning, philosophers have agonised over the question of what makes us human. Is there a difference in kind or merely a difference in degree between ourselves and other animals? . . . Direct comparisons between people and animals are often seen as demeaning, even offensive.

But such comparisons are not new, even within theological circles. Pascal wrote:

It is dangerous to show a man too clearly how much he resembles the beast, without at the same time showing him his greatness. It is also dangerous to allow him too clear a vision of his greatness without his baseness. It is even more dangerous to leave him in ignorance of both. (Pascal [1670] 1958, 418)

Evolutionary psychology can certainly help to reduce that ignorance. Today, some psychologists are so enthused by evolutionary psychology that they see it as mounting a takeover bid for the whole of psychology. For example, when David Buss published his book *Evolutionary Psychology*, he added the subtitle *The New Science of the Mind*. He then proceeded to reorganize the whole of psychology within an overall framework of evolutionary psychology (Buss 1998). Others are more modest in their claims. For example, courses in the Open University in Britain are content simply to note that "evolutionary psychology focuses on how human beings came to be the apparently special animal we are today."

## But What about Human Uniqueness?

Each phylum has properties and abilities none others do. Birds fly, but humans cannot. Humans can write poetry, but birds cannot. Is there anything in evolutionary psychology that seeks to deny the uniqueness of humans, which, as Christians claim, comes from God's gracious invitation to form a personal relationship? We believe not. Evolutionary psychology has no interest in such a question. That is not to say that individual psychologists will not have personal beliefs about such issues. As we shall see later, they will and they do. One intensively researched topic in contemporary evolutionary psychology, "theory of mind," helps to illustrate the point.

## THEORY OF MIND AND HUMAN UNIQUENESS

### What Is a "Theory of Mind"?

John Tooby and Leda Cosmides have defined *evolutionary psychology* as "psychology informed by the fact that *the inherited architecture of the human mind is the product of the evolutionary process*" (Tooby and Cosmides 1992, 7, emphasis added). Thus, the focus of research in evolutionary psychology is the question of how humans came to possess the characteristics and capabilities that we see in the human population today. One of those characteristics is the ability to make inferences about what is going on in another person's mind (Byrne and Whiten 1988).

As far back as 1978, David Premack and Gary Woodruff had described animals that seemed to have the ability to understand what another animal was thinking. They claimed these animals had a "theory of mind" (Premack and Woodruff 1978). According to two acknowledged leaders in the field, Andrew Whiten and Richard Byrne:

> If an individual is able to respond differentially, according to the beliefs and desires of another individual (rather than according only to the other's overt behavior), then it possesses a theory of mind. (Whiten and Byrne 1997, 8)

### Is Mind-Reading a Unique Human Ability?

Humans gain a theory of mind by the time they reach four years of age, although this ability may be diminished in people with autism spectrum disorder and related disorders. There is considerable controversy about whether nonhuman animals, even those species that display complex social behavior, have a genuine theory of mind. If the existence of a theory of mind can be predicted from brain size or brain/body ratio, the nonhuman primates should be the most likely animals to have a true theory of mind. However, as Richard Byrne points out:

> Quite what benefits a large neo-cortex brings—the underlying cognitive basis of monkey and ape social sophistication—is not straightforward to answer. . . . Researchers have to be very cautious, then, in attributing to non-human primates the ability to understand social behavior or how things work in the mechanistic way of adult humans. Rapid learning in social circumstances, a good memory for individuals and their different characteristics, and some simple genetic tendencies are capable of explaining much that has impressed observers as intelligent in simian primates. (Byrne 2006, 99)

Many Christians believe that God created humans to be distinct from, and superior to, other animals. They may be tempted, in a search for proof of human uniqueness, to seize upon mind-reading as a way of uniquely separating humans from nonhumans, thus creating a concordism between their theological assumptions and scientific research on mind-reading behavior. The danger is that, if researchers someday are able to establish conclusively that fully developed mind-reading occurs in nonhumans, this concordism will collapse. If that happens, some will argue that humans are therefore "nothing but" unusually complex primates, thus ignoring the genuine distinctiveness of the ethical, moral, and religious aspects of human cognition and behavior.

Frans de Waal, in his book *Good Natured*, cautioned against this type of naïve reductionism:

> Even if animals other than ourselves act in ways tantamount to moral behavior, their behavior does not necessarily rest on deliberations of the kind we engage in. It is hard to believe

that animals weigh their own interests against the rights of others, that they develop a view of the greater good of society or that they feel lifelong guilt about something they should not have done. . . . To communicate intentions and feelings is one thing; and to clarify what is right, and why, and what is wrong, and why, is quite something else. Animals are no moral philosophers. (de Waal 1996, 209)

In a similar vein, Richard Byrne warns against excessive attempts to attribute human characteristics to nonhuman primates:

These [behaviors] are not carefully thought out by beasts; and nor are any genes really selfish or altruistic, they are no more than pieces of DNA molecules; nor is an understanding of kinship likely to be remotely similar to our own. . . . These are human-applied labels based on the superficial appearance of the actions of individual animals whose behavior is partially governed by genes. . . . Human social behavior is influenced by our culture and our extensive information transmission by spoken and written language in ways not well described by biology. (Byrne 2006, 96)

On the other hand, the views held by evolutionary psychologists can change rapidly. Prior to 2000, Michael Tomasello was one of the researchers who argued that only humans have a "theory of mind" (Tomasello 2000). Later, Tomasello's own laboratory studies convinced him that his earlier views needed to be revised. In 2003 he wrote:

In our 1997 book *Primate Cognition* we reviewed all the available evidence and concluded that nonhuman primates understand much about the behavior of conspecifics but nothing about their psychological states, [but] . . . in the last five years new data have emerged that require modification of this hypothesis. . . . We are now convinced that at least some non-human primates—the research is mainly on chimpanzees—do understand at least some psychological states in others. . . . For the moment, we feel safe in asserting that chimpanzees can understand some psychological states in others, the question is only which ones and to what extent. (Tomasello, Call, and Hare 2003, 153, 155–56)

Thus, it would have been foolish in 2000 to concoct a concordism between the current findings of evolutionary psychology and particular interpretations of scripture, because the concordism would have been invalidated by the shifting research results.

*Does Research on Mind-Reading Threaten Christian Faith?*
Despite the concern that some Christians may have about the apparent narrowing of the gap between humans and some nonhuman primates, we believe that there are no great issues at stake in evolutionary psychology research on any topic, including theory of mind. A Christian can be enthusiastically open-minded about developments in evolutionary psychology, seeing in some of its developments new insights into what does and does not make us uniquely human and made "in the image of God." Christians who are discerning may glimpse fresh pointers to the greatness of the Creator in the wonders of the creation.

To summarize what we have been trying to convey about evolutionary psychology, we offer these points:

The study of animal behavior and animal cognition has a long history within psychology. Comparative psychologists and evolutionary psychologists have studied how learning, memory, and problem-solving in various animal species are related to the complexity of their nervous systems. Identifying the similarities and differences between the nervous systems of animals and humans was, and is, an important scientific topic in its own right, and has within it no special stakes for a Christian.

There were occasional attempts to identify differences between animals and humans that would uniquely set humans apart from non-humans. In the past, frequent candidates for these differences were language and symbolic processes. More recently, some researchers have claimed that a sophisticated aspect of social cognition called "mind-reading" is found only in humans, while other researchers believe that it can be observed in nonhuman primates.

We suggest that there are no theological stakes here, only some very interesting science.

## The Neural Basis of Mind-Reading

There are interesting links between neuropsychology and aspects of contemporary evolutionary psychology with respect to theory of mind. Two specific research findings illustrate this: the discovery of mirror neurons and von Economo neurons. These neuroscience discoveries are directly relevant to a better understanding of the evolutionary development of mind-reading (involving mirror neurons) and of some forms of animal and human social behavior (involving von Economo neurons).

### Mirror Neurons

In the late 1980s Giacomo Rizzolatti and his colleagues at the University of Parma discovered what have become labeled as *mirror neurons* (di Pellegrino et al. 1992). Initially, this discovery attracted little attention. However, it was thrust into the limelight when the high-profile neuroscientist Vilayanur Ramachandran wrote:

> I predict that mirror neurons will do for psychology what DNA did for biology: they will provide a unifying framework and help explain a host of mental abilities that have hitherto remained mysterious and inaccessible to experiments. . . . Thus I regard Rizzolatti's discovery . . . as the most important unreported story of the last decade. (Ramachandran 2000)

Mirror neurons are located in an area known as F5 in the premotor cortex of the monkey brain—an area involved in controlling complex hand movements, and analogous to Broca's area in humans (which is involved in spoken language). Mirror neurons in a monkey's brain become active when the monkey performs a goal-directed hand motion, such as picking up a peanut and bringing it to the mouth. What makes these neurons unusual is that they become equally active when that monkey simply watches another individual (a different monkey or even a human) perform a similar action (Kilner and Lemon 2013). For this reason, they were labeled by some as "monkey-see, monkey-do" cells. Vittorio Gallese, one of the coauthors of Rizzolatti's original paper, speculated that these mirror neurons may underlie the process of mind-reading, or are at least a precursor to such a process (Gallese and Goldman 1998).

As a reminder, *mind-reading* refers to the activity of representing to oneself the specific mental state of others, including their goals, their perceptions, their beliefs, and their expectations. According to Gallese and Goldman (1998), this type of mind-reading develops automatically in humans—with no special training needed. Other researchers have speculated that mirror neurons provide a neural foundation for the development of social relationships in both humans and nonhuman primates.

To return to the connection between neuroscience and evolutionary psychology, we note that evolutionary psychology argues against a sharp separation between cognitive and social capacities. The unprecedented complexity of human beings as compared to monkeys and great apes has come about precisely because cognitive and social domains are integrated in mutually reinforcing ways. Therefore, Christians are on thin ice if they claim mind-reading as a God-given, uniquely human ability, or to claim that humans have a unique capacity for social relationships based on being made in the image of God.

## Von Economo Neurons

The capacity for social relationships is, according to evolutionary theory, an evolved capacity. Is there evidence that human social skills may have taken a quantum leap when combined with new cognitive abilities—equipping *Homo sapiens* with capacities and achievements clearly different from those of our nearest primate relatives?

In *Neuroscience, Psychology, and Religion* (Jeeves and Brown 2009), we noted the recent interest in *von Economo neurons* (VENs), which are very large neurons that have unusually long axons projecting throughout much of the cerebral cortex. There is speculation that the proliferation of von Economo neurons in humans might be the critical element for the development of a "deep social mind" (Whiten 2006; Brown and Strawn 2012).

Some background may be helpful. In humans, VENs originate in the limbic system within the anterior cingulate gyrus and insula. The insular cortex receives information about the state of the body from the internal organs, including information about body responses to emotions. The anterior cingulate cortex consistently shows activity when an individual is making decisions in the social and moral domain or is experiencing social emotions. For example, brain scans reveal that the anterior cingulate cortex and the insula show enhanced activity during states of empathy, shame, trust, and humor, as well as while interpreting the mental and emotional states of other people, or while pondering moral decisions that involve others.

According to a theory proposed by neuroscientists John Allman and Patrick Hof, feedback about bodily emotions converges on the anterior cingulate and insular cortex, which then distribute this information throughout the cortex by way of the VENs. This process triggers cognitive reactions to these emotional states. The integration of bodily states with higher cognition allows the human brain to interpret the emotion and understand its social significance (Allman et al. 2010). Thus, VENs are important in any discussion about the "deep social mind" of humanity.

### Are Von Economo Neurons Unique to Humans?

Other studies from Allman's research group have reported VENs in other animals with exceptionally large brains: the great apes, several types of whales, and elephants. The location of VENs is similar in all the species that have them. They are most abundant in the frontoinsular cortex (area FI) and are also present at lower density in the anterior cingulate cortex. Additionally, they are found in a dorsolateral prefrontal area and less abundantly in the region of the frontal pole. The VEN morphology appears to have arisen independently in hominids, whales, and elephants, and may reflect a specialization for the rapid transmission of crucial social information in very large brains (Hakeem et al. 2009).

Clearly, the mere presence of von Economo neurons does not make humans unique, although the exceptionally large number of VENs in humans certainly boosts the computational power of the human brain. Remember that evolutionary psychology has tended to reject any theoretical efforts to separate cognitive capacities from social skills and experiences, because it views these two domains as integrated in mutually reinforcing ways. As we mentioned earlier, perhaps it is the combination of the two domains that allowed the complexity of human behavior to reach unprecedented levels compared to monkeys and great apes. Extensive social interactions may have been crucial in shaping the human nervous system and priming the appearance of the highest and most complex cognitive skills, including language. The discovery of von Economo neurons provides a further insight into how this may have occurred.

### Wisdom, Moral Behavior, and Von Economo Neurons

As we mentioned in chapter 8, Warren Brown and Lynn Paul have described ways in which wisdom and other aspects of moral and social behavior are compromised by brain dysfunction (Brown and Paul 2015). Given that much of the practical problem-solving associated with wisdom occurs in a social context, anything that would impair our perception and interpretation of the behavior of other people should diminish wisdom. Brown and Paul note that von Economo neurons are involved in detecting the emotional states of other people and analyzing the social significance of our own actions and emotions. VENs then integrate that information and pass it on to multiple cognitive networks, allowing the person to make a wise decision in that particular social setting. We see this as another example of the relevance of neuroscience and evolutionary psychology to a better understanding of central aspects of human behavior.

## EVOLUTIONARY PSYCHOLOGY AND THE *IMAGO DEI*

### *Human Distinctiveness as Seen by Biblical Scholars and Theologians*

The central issue underlying this chapter is human uniqueness, as investigated through scientific research that compares humans to other animals. Can what God has allowed us to discover about human-animal similarities and differences through research in evolutionary psychology offer any guidance or new insights into what does and does not make humans different? Can it help eliminate untenable claims about specific mental abilities that make humans different? As we continue to pursue that issue, in this section, we also consider evidence from biblical scholarship about what makes humans special.

For centuries, the traditional way of defining "humanness" was to seek particular criteria allegedly unique to the human condition: bipedalism, opposable thumbs, tool-making, learning ability, abstract thought, shame, play, artistic sense. In the past fifty years, biblical scholars and theologians have proposed that the Bible's distinction between humans and all other animals is based on a different criterion known as *imago Dei*—that humans are made "in the image and likeness of God" (Genesis 1:26). But what does that mean?

To illustrate the lively debate among biblical scholars on this topic, consider the perspectives described by James Barr (1993, 157), who identified five ways in which *imago Dei* has been interpreted in the past:

▸ As the capacity for rational thought—proposed by Augustine and Aquinas and accepted by Luther and many of the Reformers.

- As the possession of a "soul."
- As the possession of physically distinctive characteristics, such as bipedalism.
- In terms of agency or dominion over the world. According to Gerhard von Rad, *imago Dei* is not what we *are* but what we are *called to do* (von Rad 1961, 57).
- As the capacity for relationship—with God and with other creatures. According to Karl Barth, *imago Dei* is not just an ability for relationship, but the relationship itself, a relationship with God and with each other, most clearly exemplified in Jesus, who alone is fully the image of God (Barth 1960, 220).

Imago Dei: *Something We Possess, or Something We Are?*
What becomes clear from the views of these and other scholars is that *there is now a general agreement that the* imago Dei *is not anatomical, genetic, neurological, or behavioral*, and that it combines functional and structural elements (Blocher 1984; Middleton 2005). Chris Wright put it well:

> We should not think of the image of God as an independent "thing" that we somehow possess. God did not *give* to human beings the image of God. Rather it is a dimension of our very creation. The expression "in our image" is adverbial (that is, it describes the way God made us), not adjectival (that is, as if it simply described a quality we possess). The image of God is not so much something we possess, as *what we are*. To be human is to be the image of God. (C. Wright 2004, 119)

Thus, while evolutionary biology and evolutionary psychology assert that we are descended from apes, nevertheless scripture affirms that we differ substantively from them in "being made in God's image" through a divine and sovereign act of the Creator.

## Imago Dei *as a Covenantal Relationship*

*Imago Dei* as the capacity for a unique relationship with God is emphasized by Joel Green, who wrote, "Genesis does not define humanity in essentialist terms but in relational—more specifically, identifying the human person as Yahweh's partner" (Green 2004, 196). Anthony Thiselton agreed, identifying the capacity for a relationship with our creator God as one of the aspects of what it means to be made "in the image of God" (Thiselton 2015, 477). Claus Westermann argued for the centrality of this relationship in defining what it means to be a person: "The relationship to God is not something which is added to human existence; humans are created in such a way that their very existence is intended to be their relationship to God." As an example of this intentional relationship, in the Genesis account (Genesis 1:28), only humans, not the other animals, are addressed directly by God (Westermann 1984, 158).

Theologians Jürgen Moltmann and Wolfhart Pannenberg added an eschatological dimension to the relationship idea. A key word for Pannenberg is *exocentricity*, emphasizing that we are constantly reaching beyond our experiences of the present world in a search for fulfillment and meaning. Together with Moltmann, he believed that there is a fundamental self-transcendence that will ultimately find its proper identity only in Jesus Christ, who fulfills the image of God in its entirety.

Unfortunately, there is no consensus among biblical scholars about:

- ▶ Whether God created humans *de novo* ("brand new" and quickly), with no continuity with the rest of creation.
- ▶ Whether *imago Dei* appeared naturally as an emergent property during human development.
- ▶ Whether *imago Dei* was "implanted" at a specific time in history, as a direct action of God.

## *Insights from Evolutionary Psychology*

Given these perspectives from the book of Genesis, does studying the "book of creation" through evolutionary psychology afford any checks and balances to our theological thinking about these issues? Does it provide any fresh insights into what should be the focus of our attempts to understand in what sense humans are "wonderfully made" (Psalm 139:14)?

We have seen that biblical scholars have identified the capacity for relationship with God and God's creation as the cornerstone of human uniqueness. Interestingly, the idea that a deep form of relationality is a distinctive feature of the human species has been proposed within scientific discussions of human nature. For example, Andrew Whiten, a leader in evolutionary psychology who studies the social behavior of monkeys, has theorized that what distinguishes humankind from the rest of the animal world is a "deep social mind." His claim is that

> humans are more social—more deeply social—than any other species on earth, our closest primate relatives not excepted. . . . By "deep" I am referring to a special degree of cognitive and mental penetration between individuals. (Whiten 2006, 212)

Whiten also argues that these deep and rich forms of human interpersonal relatedness are dependent on a basic set of human mental capacities. These capacities include language, the ability to understand the mental life of other persons (a theory of mind), forethought (contemplating the social consequences of behavior), and deeper forms of emotional attunement to other persons. While chimpanzees can be shown to possess at least rudimentary forms of these capacities, each is markedly enhanced in human beings (Whiten 2006).

### Insights from the Cognitive Science of Religion

In a book titled *Cognitive Science, Religion and Theology* (Barrett 2011), Justin Barrett made a strong case for the potential relevance of a new academic specialization, called the "cognitive science of religion (CSR)," to the work of theologians. In answer to a question once posed to him by a theologian colleague, he wrote:

What does cognitive science have to tell me that I desperately need to know as a theologian? My short answer is this: cognitive science is rapidly gaining prominence in shaping how people think about themselves and the world, and a theologian who ignores it voluntarily surrenders a useful tool for the scholarly or pastoral vocation, and risks limiting her relevance to the contemporary world. (Barrett 2011, 146)

Barrett has worked with Matthew Jarvinen on what cognitive science can add to our understanding of what it means to speak of humans being made in *the image of God*. They began with the theological assumption that *imago Dei* has something to do with the human "capacity for receiving, experiencing, and being formed through God's love" (Barrett and Jarvinen 2015, 169). This relational capacity requires that humans be able to understand the desires and intentions of the other—that is, have a theory of mind. Barrett and Jarvinen proposed that nonhuman primates and human infants have a simplistic, lower-order theory of mind. What makes adult humans distinctive is the fact that they have developed a higher-order theory of mind that allows a deeper, richer relational bond with other humans and with God. According to Barrett and Jarvinen, this new cognitive capacity is what marks humans as containing the image of God.

## Concluding Cautionary Comments

We find this CSR analysis by Barrett and Jarvinen to be interesting and insightful. But we are concerned by the speculative nature of their bold claim about human uniqueness being dependent upon a "higher-order theory of mind." We saw earlier how the views held by researchers can change rapidly, as, for example, in the dramatic turnaround by Michael Tomasello concerning theory of mind in nonhuman primates. A similar fate could await Barrett and Jarvinen. Thus, to restate one of our themes, it may be unwise to tie the interpretation of *imago Dei* so tightly to one aspect of contemporary cognitive science.

Indeed, Justin Barrett has already come forward with an alternative hypothesis. In a chapter headed "*Imago Dei* and Animal Domestication," Barrett and his colleague Tyler Greenway wrote:

> Drawing on insights from cognitive and evolutionary psychology, we propose that animal domestication in particular may have been a critical point in human religious evolution, and may even be central to an understanding of the role humans play as beings created in the image of God, or *imago Dei*, as articulated in Genesis. (Barrett and Greenway 2017, 65)

Referring to Barrett's earlier claim that a "higher-order theory of mind" is the critical component of *imago Dei*, they continued:

> We place these two hypotheses side-by-side not because we are convinced that either one is clearly superior to the other in terms of available scientific evidence or theological considerations—indeed, we regard them as potentially complementary—but because they jointly illustrate the productivity in bringing cognitive and evolutionary perspectives on religion together with theology. (Barrett and Greenway 2017, 65)

This quotation illustrates two important points we have emphasized throughout. First, that it is unwise to seize upon a particular development in science and attempt to use it to support one interpretation of scripture, and second, that there are potential benefits to be gained from "bringing cognitive and evolutionary perspectives on religion together with theology"—uncovering mutual insights and enrichments.

In this chapter we have tried to show that evolutionary psychology has substantial explanatory power that must be taken into account as we consider the relationship between science and faith. However, we also recognize that there are pitfalls within some evolutionary reasoning—for example, insisting that there must be some selective genetic pressure behind every behavior. This fails to consider the enormous power of the human brain to adapt to unique environmental situations by developing a novel form of emergent behavior that confers no long-term advantage for survival or reproduction in the general population. Also, much of the theorizing in evolutionary psychology assumes that we know with certainty about the details of early human social structure during the period in which key aspects of cognitive and social tendencies supposedly evolved. This of course leaves room for making evolutionary claims based on "just-so" stories that conveniently propose an early human social environment that matches the specific characteristics one is studying.

# CHAPTER 10

# Insights about Human Needs and Motivation

IN CHAPTER 5 we noted how, in journals such as the *Journal of Psychology and Theology* and the *Journal of Psychology and Christianity*, the science-faith discussion has focused on the areas of counseling and clinical psychology. We mentioned our concern that this approach ignores issues that might arise when relating other major areas of contemporary psychology such as neuropsychology, evolutionary psychology, cognitive psychology, and social psychology to Christian faith. For that reason, we have devoted five chapters to those major topics, which figure prominently in typical university courses of psychology. The research we surveyed in those chapters provided evidence that brain function (and dysfunction)—shaped by genetic factors that were honed through natural selection and influenced by the immediate physical and social environment stimulation—must be taken into account as we consider our moral, religious, and interpersonal experiences and behaviors.

In this chapter we look briefly at human needs and human motivation. Although neither of us claims expertise on these topics, we hope that our reading of the literature is not seriously wide of the mark. In guiding our own thinking, we are indebted to Kenneth Boa for his extended treatment of this topic in his Oxford University thesis "Theological and Psychological Accounts of Human Needs: A Comparative Study" (Boa 1994). Human needs and motivations are topics of concern both to psychological scientists and biblical scholars, as well as to ordinary thoughtful Christians. *What is different about these topics is that they have received more attention from Christian psychotherapists and pastoral counselors*—the groups that have traditionally been the most interested in attempting an integration of psychology and faith. Because these topics have been covered thoroughly in some of the publications mentioned in chapters 1, 4, and 5, there is perhaps less reason for us to address them here. However, we include them to demonstrate that the theme of our book also applies to these topics: by avoiding concordisms, it is possible to identify mutual insights and enrichments for both psychologists and theologians.

## Some Cautions about Language

The Bible has many things to say about human nature and human motivation. Psalm 8, for example, reminds us of the theocentric context of scriptural statements, in that they deal primarily with humans in relation to God as Creator, Sustainer, and Redeemer. This is a timely reminder that the kinds of questions posed by twenty-first-century scientists were not even considered by the biblical authors, let alone answered by them. In order to understand some of the key features of the biblical portraits of humans in relation to portraits from contemporary science, we need to remind ourselves of the timely advice given by John Stott in his commentary on Paul's letter to the Romans. Commenting on the seventh chapter of Romans, Stott wrote:

> It is never wise to bring to a passage of Scripture our own ready-made agenda, insisting that it answers *our* questions and addresses *our* concerns. For that is to dictate to Scripture instead of listening to it. We must lay aside our presuppositions, so that we can consciously think ourselves back into the historical and cultural settings of the text. Then we shall be in a better position to let the author say what he does say and not force him to say what we want him to say. (Stott 1994, 189)

Insights into human nature are found within the world's literature, portrayed in its art, and proclaimed by different religions. These insights long predate any research findings offered by scientists. We must be careful not to be too critical about some of the earlier interpreters of scripture, who did not have access to the full textual material and the relevant historical context, and thus had no doubt that it was proper to interpret certain passages literally. For example, early theologians believed that the earth

- ► Rests on pillars (1 Samuel 2:8)
- ► Does not move, but is stationary (1 Chronicles 16:30)
- ► Has ends and edges (Job 37:3)
- ► Has four corners (Isaiah 11:12; Revelation 7:1)

The key point here is that *improved geological and astronomical knowledge about the earth and the universe has dramatically altered the way that contemporary theologians interpret passages such as the ones listed above.* Theologians today do not see any conflict between the Bible's poetic descriptions of the "pillars of the earth" and scientific descriptions of the earth's position in the solar system.

This principle should apply just as strongly to the *interpretation of scriptural passages about human nature in the light of psychological science.* The biblical statements about human nature and personality are embedded in the forms of thought and cultural assumptions of the time in which they were written. We must avoid the danger of directly comparing biblical statements about human nature with descriptions of human characteristics derived from psychological science. Those direct comparisons would likely generate false conflicts—thereby repeating the mistakes of those who, in the time of Galileo, wrongly applied the emerging evidence from astronomy to biblical statements about the earth and the stars.

The temptation to do such direct comparisons is greater in psychology than in the physical sciences. The language used in the two contexts— biblical and psychological—at times looks highly similar, especially when talking about human needs, relationships, fears, and aspirations. The pronouncements of some psychotherapists, for example, sound remarkably similar to statements we find about human motivation and human nature in the Bible. It is not always clear that the scriptural language may have meant something quite different from the meaning of the words used by psychologists today. As Gordon Allport noted many years ago:

> The religious vocabulary seems dignified but archaic; our scientific vocabulary, persuasive but barbaric. "His id and superego have not learned to cooperate," writes the modern mental hygienist; "The flesh lusted against the Spirit, and the Spirit against the flesh," writes Paul. "Feelings of guilt suggest poor personality teamwork," says the twentieth-century specialist; "Purify your hearts, you double minded," exhorts James. "The capacity of the ego to ward off anxiety is enlarged if the ego has considerable affection for his fellows and a positive goal to help them." Correspondingly, St. John writes, "Perfect love casts out fear." It would be difficult, I suspect, to find any proposition in modern mental hygiene that has not been expressed with venerable symbols in some portions of the world's religious literature. (Allport 1950, 86)

By contrast, the language found in today's specialist journals in cognitive science, neuropsychology, and evolutionary psychology is completely different from anything we might find in scripture—almost as foreign as the language of nuclear physics or genetics or molecular biology. So the temptation to find science-faith conflicts within these areas or to seek concordisms is not so strong as in the case of personality theory and counseling.

It is important to keep these concerns about language in mind as we turn to some topics within psychology that have direct bearing on issues central to Christian belief and Christian living.

## PSYCHOLOGICAL PERSPECTIVES ON HUMAN NEEDS AND MOTIVES

Why do people do the things that they do? What drives them to commit acts of hatred and violence, or acts of compassion and love? What needs are they seeking to meet? Theologians and philosophers have pondered these questions for centuries, and for the past century psychologists have joined in the discussion.

Theories of personality contain basic assumptions and claims about human nature. Each theory presents a model composed of various hypothetical constructs or entities that are said to interact in specified ways, both within the individual and with the environment. The outcome of such interactions guides our thoughts, feelings, and behavior. All such models aim at increasing our understanding of optimal, as well as dysfunctional, behavior and mental processes. Not surprisingly, since all these models attempt to explain the same events, they do possess certain common features and share certain pervasive themes. One such theme involves identifying common human needs, locating their roots, and investigating what goes wrong when such needs are unfulfilled. The path to satisfying those needs might be blocked by other forces within the personality (according to the intra-psychic models), or by the demands of the external social and cultural environment (according to the psychosocial models). Our aim here is to consider how we might relate the descriptions of human nature proposed by five prominent personality theorists to those derived by theologians from scripture.

### Freud (1856–1939) and Erikson (1902–1994)
Sigmund Freud's *psychoanalytic theory* proposed that the personality consists of three components—*id*, *ego*, and *superego*—along with a small number of basic psychic processes whose mutual interactions shaped the personality and mental health of an individual. Freud wrote, "The forces which we assumed to exist behind the tensions caused by the needs of the id are called instincts" (Freud 1949, 5). He saw the strident demands

of the id as arising from innate drives for survival and reproduction. These sexual and aggressive drives are opposed by the superego's moral standards, derived from expectations of parents and society. The ego, as the executive part of the personality, must somehow reconcile those opposing forces—satisfying the id's drives without offending the superego—in order to accomplish the tasks of daily life in the real world. Freud claimed that the way those internal, unconscious conflicts were handled would determine the mental life and interpersonal relationships of that individual.

Several decades later, Freud's model was extended by one of his disciples, Erik Erikson, a very influential figure in the history of *psychoanalysis*, as Freud's theory came to be called. Erikson originally worked within the standard Freudian framework, which emphasized the unconscious conflicts within an individual's personality, often entangled in the childhood relationship with the parents. However, Erikson gradually moved toward a "life cycle" view of personality as developing across eight chronological periods in a person's life (Erikson 1950). Within each period, Erikson identified a specific need or challenge that must be mastered at each stage. The first four stages cover infancy and childhood, as the infant first forms a special bond with the parents (developing *trust*), then spends the childhood years gradually disengaging from the parents (developing *autonomy* and *initiative*), then building a sense of individual *competence*, to prepare for an independent life in adulthood.

During adolescence, the person struggles to achieve a sense of *identity*. Who am I? What is important to me? Where am I headed in life? In many cases, career choice flows out of identity formation (Erikson 1968). The related question of *intimacy* soon arises. With whom do I want to share my life? The choice of a mate is built on the successful resolution of earlier challenges, meeting the earlier needs. Specifically, a mature capacity for intimacy is built on the foundation of a secure sense of *identity*.

What is the next step for the adult who has achieved identity and intimacy? Erikson claims that mature adults struggle to express their *generativity*—a concern for the next generation and for the future of society. Those who focus on their own needs and achievements will eventually stagnate, while those who turn outward to make a difference in other people's lives will find their own lives renewed with vigor and a new sense of purpose. Finally, in late adulthood, Erikson saw a final challenge: a need to develop a sense of *ego integrity*—to be able to look back on life and see that the various events actually fit together to form a meaningful and worthwhile whole.

## Maslow (1908–1970) and Rogers (1902–1987)

Two other well-known twentieth-century personality theorists were Abraham Maslow and Carl Rogers. They've often been described as the architects of the "humanistic movement" in modern psychology. Unlike Freud, who emphasized the antisocial, "evil" aspects of human motivation, Maslow believed that human nature is essentially good. Left to themselves, individuals will naturally move in the direction of reaching their full potential—which Maslow called *self-actualization* (Maslow 1968). He proposed that humanity shares certain basic motives that can be arranged into a "hierarchy of needs." The needs at lower levels (*physiological needs* and *safety needs*) must be addressed and satisfied before those at higher levels (*affiliation needs* and *achievement needs*) begin to dominate a person's motivation. Maslow explicitly considered the relationship between religion and his theory (as part of the broader humanistic psychology movement that he founded). He proposed that humanistic psychology could serve as a secular surrogate for religion. He wrote:

> The human being needs a framework of values, a philosophy of life, religion or a religion-surrogate to live by and understand by, in about the same sense that he needs sunlight, calcium or love. Without the transcendent and the transpersonal, we get sick, violent, and nihilistic, or else hopeless and apathetic. We need something "bigger than we are" to be awed by and to

commit ourselves to in the new, naturalistic, empirical, non-churchly sense. (Maslow 1968, iv)

Carl Rogers incorporated Maslow's ideas—about the positive potential of humans and the movement toward self-actualization—as the basis for his new method of psychotherapy, which he called "client-centered therapy" to distinguish it from other therapeutic approaches (Rogers 1951). According to Rogers, the goal of therapy was to create the necessary conditions to enable an individual to utilize and express the inner growth-directed tendency that every person possesses. Thus, a client voluntarily seeks help for a problem, but does not relinquish responsibility for the overall situation, because clients possess within themselves the power to overcome their own psychological problems (Rogers 1957). Like Maslow, Rogers progressed from presenting a psychological model to developing a philosophy of life. In his book *A Way of Being*, he wrote:

> I am no longer talking simply about psychotherapy, but about a point of view, a philosophy, an approach to life, a way of being, which fits any situation in which growth—of a person, a group, or a community—is part of the goal. (Rogers 1980, ix)

Like Maslow, Rogers believed that there was no evil within a person, and therefore no genuine guilt. External judgments or criticisms are not valid or helpful, so a therapist should strive to maintain a nonjudgmental "unconditional positive regard" toward the client. A key point to notice here is how a psychological model can, at first imperceptibly, and then explicitly, develop into a philosophy of life with moral and ethical overtones.

### *Fromm (1900–1980)*

Our final example is Erich Fromm, who was strongly influenced by Sigmund Freud and Karl Marx, as well as by ideas from the Orthodox Judaism of his childhood. He took Freud's intrapersonal dynamics and sought to apply them to a deeper psychological understanding of groups and societies. He believed that human personality can be understood as the coexistence of animal qualities and human characteristics. While the animal aspect of human nature produced physiological drives such as hunger, thirst, and sex, other uniquely human needs also must be satisfied to achieve true happiness. These needs, Fromm believed, included caring relationships; a sense of identity, freedom, and independence; and actively striving for and accomplishing worthwhile goals (Fromm 1964). In the modern world, human nature is being shaped by the economic and social structures in which people live, but is still constrained by these basic psychological needs. The important point here is that Fromm made the presence and the fulfillment of human needs a crucial part of his psychological theory.

### Psychological Needs: Taking Stock

Our very brief survey makes it obvious that the list of psychological needs suggested by these different personality theorists contain common themes, such as the idea that human motivation goes well beyond the drives based on individual survival, that a sense of identity is important, and that optimal functioning requires both a sense of social connection and the freedom to make independent choices. But these theories contain remarkably divergent perspectives on issues such as whether people are basically good or evil, whether human motivation is primarily conscious or unconscious, and whether tension within the personality helps or hinders personal growth. How are we to decide which, if any, of these lists of needs to accept? The difficulty in deciding among them should certainly trouble anyone claiming a scientific basis for these theories. If personality psychologists want to attach the label "scientific" to their models, they would need to provide evidence showing how well the models fit the established research findings about typical patterns of adaptive and dysfunctional behavior.

Given that all of these theories attempt to explain the same reality, how can we explain their diverging assessments of human nature? For example, Freud believed that human nature is essentially negative; he was distrustful of people and pessimistic about their intentions. In contrast, Maslow and Rogers believed that human nature is primarily positive and tends toward self-actualization and fulfillment; they thought Freud's perspective was biased as a result of concentrating on individuals with psychopathology. As we will see, theologians and biblical scholars across many centuries have also taken positions on human nature and human needs. How should theologians' accounts relate to those of psychologists?

## THEOLOGICAL PERSPECTIVES ON HUMAN NEEDS AND MOTIVES

We select four examples, spanning a millennium and a half of church history, to illustrate how leading theologians have written about human needs.

### St. Augustine (354–430): God Alone Can Satisfy

St. Augustine, widely regarded as the greatest of the Latin church fathers, believed that every human being's predicament before a holy God resulted in universal basic needs that only God could satisfy. Augustine contrasted the depths to which humanity has fallen through sin with the heights to which it can be raised by the redemptive grace of God. This theme permeated much of Augustine's writings and has been described as "the cornerstone of Augustinian anthropology" (Sullivan 1963). For example, in his *Confessions* we read about his own struggles that still resonate with life in the twenty-first century. Augustine had no doubt that without the direct intervention of God's grace, humans had no hope of redemption. Obviously, Augustine's perspective was very different from that of the "self-actualization" or "self-fulfillment" humanistic psychologists mentioned above. This is underlined when one reads of Augustine's profound analysis of the impossibility of any human being ever attaining happiness apart from God. He wrote:

The simple truth is that the bond of a common nature makes all human beings one. Nevertheless, each individual in this community is driven by his passions to pursue his private purposes. Unfortunately, the objects of these purposes are such that no one person (let alone the world community) can ever be wholly satisfied. The reason for this is that nothing but Absolute Being can satisfy human nature. (Augustine, *City of God* 18.2)

Not that Augustine was advocating an ascetic mentality toward the things of this world. According to him, enjoying the blessings of this world is fine, as long as they do not lure the heart away from the highest calling of eternal blessings.

## Thomas Aquinas (circa 1225–1274): Seek God's Comfort

Our second example is another major figure who exerted an enduring influence on Western Christendom, namely Thomas Aquinas. He sought to systematize aspects of Aristotelian philosophy with Christian theology. Aquinas frequently referred to the notion of human needs giving rise to motivation for actions, with an overarching goal of achieving "happiness" (in the form of fulfillment or well-being). Following Augustine, Aquinas believed that humans were not capable of achieving complete happiness without the presence of God in their life.

According to Thomas, because we all have an inherited tendency to sin, humans are basically evil, in the sense that we are constantly in opposition to God's will. Thus, as Étienne Gilson has observed, "At the basis of Aquinas's philosophy, as at the basis of all Christian philosophy, there is a deep awareness of wretchedness and need for a comforter who can only be God" (Gilson 1957, 375). We need God's comfort to relieve our fears and anxieties, and we need God's grace to restore the good within us so that we can live according to God's will.

*Jonathan Edwards (1703–1758):*
*Restore a Broken Relationship*

Our third example moves to a different era, a different tradition, and a very different theological position. The writings of Jonathan Edwards were largely shaped by his religious experiences. In his view, the themes of God's sovereignty, holiness, and grace stand alongside those of human sinfulness and the need for redemption. Steeped in the older Puritan authors, his views illustrate key features of the Reformed (Calvinist) tradition with regard to human needs. He certainly recognized and taught the need for a basic change in the condition of the human heart by the power of the Spirit of God. Indeed, the transformation that he saw as necessary was as radical as that which takes place in going from death to life. He wrote:

> Affections that are truly spiritual and gracious, do arise from those influences and operations on the heart, which are spiritual, supernatural, and divine. . . . And the influences of the Spirit of God in this . . . [to] communicate himself, and make the creature partaker of the divine nature, this is what I mean when I say that "truly gracious affections do arise from those influences that are spiritual and divine." . . . And natural men are represented in the sacred writings as having no spiritual life, no spiritual being; and therefore conversion is often compared to opening of the eyes of the blind, raising the dead, and a work of creation. . . . It is grace that is the seed of glory and dawning of glory in the heart, and therefore it is grace that is the earnest of the future inheritance. (Edwards 1821, 128, 135, 136, 175)

Thus, the human needs of which Edwards wrote were the fundamental need for the restoration of the broken relationship with God and the need to place all one's affections and hopes in God's promises of a future inheritance in Christ.

## Karl Rahner (1904–1984):
### Embark on a Transcendent Relationship

Our final example of prominent theologians' views on human needs comes from the twentieth century and yet another theological tradition, the Catholic Church. Karl Rahner, who has been called "the father of the Catholic Church in the twentieth century," set forth an approach that has been described as transcendental-existential. He saw the basic freedom of a person as the ability to move toward the love of God or away from it. Humans do have biological needs, but because humans are more than biological, *they have needs that transcend the physical—needs that human effort cannot satisfy.*

For Rahner, a person is "always exiled in the world and is always beyond it" (Rahner 1990, 406–7). Paradoxically, for Rahner, a person "does not possess within himself what he essentially needs in order to be himself," because at a fundamental level the universal human need is to have a relationship with God.

Rahner stressed that the approach to God must be for God's own sake; God cannot be shaped to fit our needs (Rahner 1990, 2). As British theologian John Stott wrote, "We cheapen the gospel if we represent it as a deliverance only from unhappiness, fear, guilt and other felt needs" (Stott 1994, 88). According to Rahner, the attempt to reduce God to a mere fulfillment of human needs is "the unique heresy of our time" (Rahner 1990, 27). The difference between Rahner's concept of human needs and the way that personality psychologists define the concept should certainly warn us against any attempt to integrate the two within a single theoretical formulation.

## Relating Theological and Psychological Accounts

### Recognizing Complementarity of Purpose

The four theologians mentioned above represent a wide range of historical periods, theological traditions, and philosophical perspectives. They differed in significant ways from one another. All, however, in their own ways have profound things to say about the *deepest human needs*. Each perspective underlines and expands the New Testament accounts of the *universal need* for forgiveness and grace, for love of God and neighbor, and for that eternal sense of purpose and sure hope that only God can give.

The psychologist who is also Christian knows that telling the whole story about human nature and about our deepest needs must go beyond the various psychological accounts of human needs, whether those offered by Freud, Maslow, Rogers, or Fromm. These only tell the psychological "immanent" story and must be *complemented* by the theological "transcendent" story given by revelation. But that does not mean trying cleverly to harmonize the two stories or mixing bits of one with bits of the other. Such an attempt misunderstands the nature of psychological theories while simultaneously abusing scripture.

Thus, to take Thomas Aquinas as one example, while he undoubtedly provides many rich and profound insights into human nature, these insights are not organized in the categories of modern personality theories. Led astray perhaps by the use of similar words, by similar references to desires and needs, it would be *all too easy to try to integrate or incorporate* Aquinas's wisdom into psychological theories, or vice versa. The outcome of any such attempt is likely to be a muddle that does justice to neither and hence reduces rather than increases our overall understanding of human nature and human needs. Rather, we should gratefully and thoughtfully accept the enrichment that insights from the theological, transcendental perspectives can provide to those from a psychological, immanent perspective, and vice versa. Each has something to offer to the other, as long as both groups respect the work of the other, and both groups recognize the differences in purpose and audience between the two types of perspectives. To assume congruence between the accounts of the theologians and the psychologists because of some surface similarities in the language is a prescription for confusion. It is one thing to recognize that analysis in terms of psychological models may enrich a theological understanding of human nature, and vice versa. But it is quite another thing to try to interweave and integrate the psychological and theological insights into a composite model that amounts to mixing up the two in a quite uncritical and unjustified way.

### Explaining vs. Explaining Away

Karl Rahner perceptively identified important issues in relating theological and psychological accounts of human nature. He noted that psychology may indeed offer many insights on what it is to be human, but it cannot *define* the human person, nor can the mystery of what it is to be a person be reduced to psychological concepts, however sophisticated and refined (Rahner 1990). For that reason, he criticized psychological attempts to "explain away" guilt. While he acknowledged that some forms of psychotherapy may relieve a measure of suffering, it is only through God that people can be delivered from real guilt.

Psychologist Paul Vitz exposed clearly how some psychologists shift from the scientific to the philosophical and move, without justification, to theories of moral obligation. He argued that in doing so, they blur the conceptual boundaries between science and the broader views of humankind already richly found in literature, philosophy, and religion down through the centuries (Vitz 1977). On the other hand, some psychologists are either unaware or unwilling to expose their own metaphysical and moral presuppositions. This can be a problem, especially when they seek to invoke what they believe is the authority of science to authenticate their opinions.

### Models of Intellectual Partnership

We now return to our earlier central question of how properly to relate the psychological and theological accounts of human needs so that the benefits of a potentially enriched partnership can be realized. Should it be, as Stanton Jones has argued, by "incorporation" of psychological findings into a theological worldview? Should it be "closer integration" of psychotherapeutic theory and religious beliefs? What would incorporation or integration look like in this context?

For example, in the case of Freud's psychoanalytic theory, would we add a component labeled "God" to id, ego, superego, and the other hypothetical constructs? Or, in the case of Erikson, would we add a religious component at each of the eight developmental stages he identified? Would we build upon Maslow's hierarchy of needs and add another level that included religious needs? We believe that this approach to integration would be laughable, and we think Jones would agree. Jones warns against intermixing language from different domains and points out that it can lead to nothing but confusion. To insert religious aspects into a psychological theory as an "afterthought" would be the type of error Austin Farrer warned against—invoking the Holy Spirit as an explanatory concept to fit into a psychological account we have constructed. He wrote, "The Holy Spirit is God. He is not to be fitted into any psychological, or for that matter personal, explanation we choose to give" (Farrer 1964, 12). For a believer, God is not a hypothetical construct, but instead is the one who upholds and sustains all things at all times (Hebrews 1:3). This must never be forgotten as we think about the relationship between scientific, psychological accounts of behavior and therapy and those held in the religious domain. We believe that in seeking to understand human nature, we can thankfully acknowledge how the psychological accounts provide fresh insights to enrich those from theology as each makes its distinctive contribution, albeit *from different perspectives.*

# CHAPTER 11

# Social Psychology and Faith

## Stories of Conflict, Concordism, and Authentic Congruence

DAVID G. MYERS

IN THIS BOOK, my friends Malcolm Jeeves and Thomas Ludwig offer a timely message: in the space between science–religion conflict (mutual disparagement) and concordisms (findings that seemingly confirm cherry-picked biblical texts), there is a place for authentic dialogue—a dialogue between the emerging insights of science and biblical scholarship.

Our mandate for such free-spirited, truth-seeking science-religion dialogue follows from our theism, which assumes:

There is a God.

It's not us.

As Christians, we assume that humans have dignity but not deity— that we are finite, fallible creatures. If that is so, then our surest belief can be that some of our beliefs are not correct. Said differently, we presume that we are the finite creatures of an infinite God. This theism mandates humility. It compels us to hold our untested beliefs tentatively. It frees us to assess others' beliefs with open-minded skepticism. And, for matters that are amenable to scientific exploration, it encourages us to winnow truth from error through careful observation and experimentation.

Such faith-based humility and skepticism fed the roots of modern science. Even today, my own "Reformed and ever-reforming" Christian heritage encourages our participation in free-spirited scientific inquiry. Thus, we submit our tentative ideas to the test. If they survive, our confidence grows. If they collide with a wall of evidence, we rethink or revise them. That's the empirical spirit occasionally commended by scripture:

- ▸ Moses (Deuteronomy 18:22 NRV): "If a prophet speaks in the name of the LORD but the thing does not take place or prove true, it is a word that the LORD has not spoken."
- ▸ Jeremiah (28:9 NRSV): "As for the prophet who prophesies peace, when the word of that prophet comes true, then it will be known that the LORD has truly sent the prophet."
- ▸ St. Paul: (1 Thessalonians 5:21 NRSV): "Test everything; hold fast to what is good."

As I explain in a chapter in Eric Priest's *Reason and Wonder: Why Science and Faith Need Each Other* (Priest 2015):

Such ever-reforming empiricism has many times changed my mind, leading me now to conclude that parenting practices have but modest effects on children's later personalities and intelligence; that crude-seeming electroconvulsive therapy can often relieve intractable depression; that the automatic unconscious mind dwarfs the conscious mind; that traumatic experiences rarely get repressed; and that sexual orientation is a natural, enduring disposition (not a moral choice).

Faith-supported scientific inquiry also has led me to *disbelieve* certain spiritualist claims ranging from aura readings to out-of-body "frequent flyer programs." If, for example, aura-readers really can detect auras above a person's head, then they should be able to guess the person's location while seated behind a screen. If they can do so, then so much the better for their claims. If not (as seems the case), let's consider the claim discounted. (Myers 2015, 127)

For Christians, the consistent failures to confirm such paranormal claims confirm the distinction between deity and humanity. We assume we are not little gods with powers of *omniscience* (reading minds, foretelling the future), *omnipresence* (traveling out of body), and *omnipotence* (levitating objects or eradicating tumors with our mental powers). As Isaiah 46:9 KJV records, "I am God; there is none like me."

Ergo, many of us Christians in psychology seek to explore God's human creation with a spirit of curiosity and humility. Was John Calvin right to suppose that "in everything we deal with God" (Calvin 1960, 1.17.2)? If so, we should worship God with our minds as we search the natural world through scientific inquiry, seeking to discern its truths. And as we pursue an authentic convergence of scientific and religious wisdom, we can tolerate seeming conflicts, without resorting to over-simplified concordisms.

## PRAYER RESEARCH: AN EPISODE OF CONFLICT?

### *The Power of Intercessory Prayer*

When Christians offer prayers to God on behalf of someone who is suffering, they may expect that God will hear those prayers and will intervene to reduce the person's suffering. After all, according to James 5:15 (NIV), "The prayer offered in faith will make the sick person well." Back in the 1800s, Charles Darwin's cousin Francis Galton (1872) was one of the first to consider testing the effect of intercessory prayer on the length of a person's life. Do prayed-for people live longer? As an indirect answer to that question, he collected age-of-death data on people who were the objects of much prayer (such as clergy and royalty). Alas, they lived no longer than average.

### *A Concordism: Prayer Works*

Galton's longevity study was not the last word on the subject. A hundred years later, Randolph Byrd tested prayer directly in a famous study,

"Positive Therapeutic Effects of Intercessory Prayer in a Coronary Care Unit Population" (Byrd 1988). This study found that hospitalized heart patients whose first names and diagnoses were randomly assigned to three "born-again" intercessors did better than other patients on six outcome measures, including the need for antibiotics and the likelihood of experiencing congestive heart failure or a heart attack.

Believing in the power of intercessory prayer, Christians welcomed these results as evidence that seemingly supported their favorite prayer proof text: "Ask and it will be given to you" (Matthew 7:7 NIV). The Byrd study was featured in a 1997 *Christianity Today* cover article that described "how physicians and scientists are discovering the *healing power* of prayer" (Thomas 1997). The article quoted physician and prayer researcher Dale Matthews: "Prayer is good for you. The medical effects of faith on health are not a matter of faith, but of science." Intercessory prayer works, the article assured us—as a treatment seemingly comparable to an effective drug. This claim would come as no surprise to the more than 80 percent of American believers who, when surveyed, agreed that "God answers prayers" (Woodward 1997; J. Jones 2010).

### A Discord: Prayer Doesn't Work

This seeming concord between biblical statements about the power of prayer and confirming medical evidence triggered a new series of experiments that asked: Do prayers, indeed, produce cures (beyond a placebo effect)? Do prayed-for people recover faster? Are they less likely to die after a heart attack or a risky surgical procedure?

So, these experiments put prayer to the test, again and again. The results showed a consistent pattern, which I summarized in *A Friendly Letter to Skeptics and Atheists* (Myers 2008):

> ► The 1988 coronary care study—mentioned previously as showing a "prayer effect" for six outcome measures—found no difference between prayer and no-prayer groups for twenty *other* outcome measures, including length of hospital stay and likelihood of dying.

- A 1997 experiment, "Intercessory Prayer in the Treatment of Alcohol Abuse and Dependence," found no measurable effect of intercessory prayer.
- A 1998 experiment with arthritis patients found no significant effect of prayers offered by people praying from a distance.
- A 1999 study of 990 coronary care patients—who were unaware of the study—reported about 10 percent fewer complications overall for the half who received prayers "for a speedy recovery with no complications." But for major complications such as cardiac arrest, hypertension, and pneumonia, there was no difference between the prayer and no-prayer groups. The median hospital stay was the same 4.0 days for both groups.
- A 2001 Mayo Clinic study of 799 coronary care patients offered a simple result: "As delivered in this study, intercessory prayer had no significant effect on medical outcomes."
- A 2005 Duke University study of 848 coronary patients found no significant difference in clinical outcomes between those prayed for and those not.
- And then in 2006 came the likely coup de grâce for intercessory prayer as a "medical treatment," with the published result of the mother of all prayer experiments—a $1.4 million Templeton Foundation–funded comparison of more than 1,800 consenting coronary bypass patients (Benson et al. 2006). The patients had been assigned to one of three groups: one that knew that it was being prayed for by volunteer intercessors, one that did not know whether it was being prayed for (but was), and a third group that did not know whether it was being prayed for (and wasn't). The simple result: *intercessory prayer per se had no effect on recovery.*

## A Christian Perspective on Prayer

As Jeeves and Ludwig have emphasized, prematurely attaching faith claims to an early scientific result may later backfire. In this case, some of us (including Malcolm Jeeves)—as Templeton advisers who had cautioned against this study—were unsurprised. Before the experiment began, I filed a notarized statement, "Why People of Faith Can Expect Null Effects in the Harvard Prayer Experiment" (available at david myers.org). It suggested why "my understanding of God and God's relation to the created world would be more challenged by positive than negative results":

1. *The prayer concept being tested is more akin to magic than to a biblical understanding of prayer to an omniscient and sovereign God.* In the biblical view, God underlies the whole creation. God is not some little spiritual factor that occasionally deflects nature's course, but the ground of all being. God works not in the gaps of what we don't yet understand, but in and through nature, including the healing ministries that led people of faith to spread medicine and hospitals worldwide. Thus, while our Lord's model prayer welcomes acknowledging our dependence on God for our basic necessities ("our daily bread"), it does not view God as a celestial vending machine, whose levers we pull with our prayers. Indeed, would the all-wise, all-knowing, all-loving God of the Bible be uninformed or uncaring apart from our prayers? Does presuming that we creatures can "pull God's strings" not violate biblical admonitions to humbly recognize our place as finite creatures of the infinite God? No wonder we are counseled to offer prayers of adoration, praise, confession, thanksgiving, dedication, and meditation, as well as to ask for what shall (spiritually, if not materially) be given.

2. *Even for those who believe that God intervenes in response to our prayers, there are practical reasons for expecting null effects.*

**The "noise" factor.** Given that 95 percent of Americans express belief in God, most patients undergoing cardiac bypass surgery will already be receiving prayer—by spouses, children, siblings, friends, colleagues, and fellow believers and/or congregants—if not offering prayer themselves. Are these fervent prayers a mere "noise factor" above which the signal of additional prayers may rouse God? Does God follow a dose-response curve—more prayers, more response? Does God count votes? Are the pleading, earnest prayers of patients and those who love them not sufficiently persuasive (if God needs to be informed or persuaded of our needs)? Are the distant prayers of strangers participating in an experiment additionally needed?

**The doubt factor.** To be sure, some Christians believe that prayers, uttered in believing faith, are potent. But are there any people of faith who also believe that prayers called forth by a doubting (open-minded, testing) scientist will be similarly effective?

**"God is not mocked."** As Christians recalled during the 1872 British prayer test controversy, Jesus declared in response to one of his temptations that we ought not put "God to the test." Reflecting on a proposal to test prayers for randomly selected preterm babies, Keith Stewart Thomson questioned "whether all such experiments come close to blasphemy. If the health outcomes of the prayed-for subjects turn out to be significantly better than for the others, the experimenter will have set up a situation in which God has, as it were, been made to show his (or her) hand" (Thomson 1996, 534). C. S. Lewis observed, regarding any effort to prove prayer, that the "impossibility of empirical proof is a spiritual necessity" lest a person begin to "feel like a magician" (C. S. Lewis 1947, 215). Indeed, if a prayer experiment were to show that the number of people praying or the total number of prayers matter—that distant strangers' prayers boost recovery chances—might rich people not want, in hopes of gaining God's attention, to pay others who will pray for them?

3. *The evidence of history suggests that the prayers of finite humans do not manipulate an infinite God.* If they could and did, how many droughts, floods, hurricanes, and plagues would have been averted? How many stillborn infants or children with disabilities would have been born healthy? And consider the Bible's own evidence: How should the unanswered prayers of Job, Paul, and even Jesus inform our theology of prayer? If the rain falls on my picnic, does it mean I pray with too little faith, or that the rain falls both on those who believe and those who don't? Should we pray to God as manipulative adolescents—or as dependent preschoolers, whose loving parents, already knowing their children's needs, welcome the intimacy?

In the biblical view, God is not a magical genie or a celestial Santa Claus whom we call forth with our prayers. God is rather the creator and sustainer of all that is. The Lord's Prayer, which I pray daily, does not attempt to cajole a reluctant God. Rather, by affirming God's nature and our human dependence even for daily bread, it prepares us to receive what God is already providing. I can approach God as a small child might approach a loving parent who already knows the child's needs. Through such prayer, Christians express their praise and gratitude, confess their sin, voice their concerns and desires, open themselves to the Spirit, and seek the peace and grace to live as God's own people.

### New Experiments on a Real Power of Prayer

Although the seeming concordism between "Ask and it will be given" and an early prayer experiment imploded, a different set of prayer experiments substantiated some remarkable benefits of prayer, *at least for the one who prays*. In eighteen published articles, Florida State University psychologist Frank Fincham described experiments that randomly assigned religious believers either (a) to pray for someone they cared about or (b) to engage in some other specified activity, such as thinking positive thoughts about the person.

Here is a sample prayer instruction from these experiments:

> Please read the example prayer below to get an idea of the type of prayer we would like you to pray on behalf of your partner:
>
> Dear Lord, thank you for all the things that are going well in my life and in my relationship. Please continue to protect and guide my partner, providing strength and direction every day. I know you are the source of all good things. Please bring those good things to my partner and make me a blessing in my partner's life. Amen.
>
> Now, please generate your own prayer in your own words on behalf of the well-being of your partner and in the space below write a short description about what you prayed for.

Fincham, who is a prolific, award-winning research psychologist (winner of the British Psychological Society's President's Award), reported that, compared to those in a control condition, those engaged in sustained, partner-focused petitionary prayer became significantly

- More forgiving
- Healthier, with lower heart rate and blood pressure
- Improved in their cardiovascular functioning
- Less alcohol consuming
- More cooperative
- Less likely to cheat on their partner
- Happier in their marriages

As one female participant explained, "We specifically pray about matters that affect our marriage. What I find is that when I sense a disagreement brewing, I ask for guidance up front rather than afterwards. That is the measure of my growth in using prayer in my marriage." "At the time we were in the program, we were separated and facing divorce," explained a male participant. "Hesitantly, we started with little hope of any reconciliation. . . . I started to use the prayer cards and begin to evaluate myself and where our marriage went wrong. I can now report that we are *not* getting a divorce but we are planning to renew our vows. Prayer definitely works even if the other partner isn't praying for the marriage" (Fincham 2013, 347–48).

To what should we attribute these benefits of praying for a partner? Through prayer, said Fincham, we become one with our partner "vis-à-vis Deity" (Fincham, Lambert, and Beach 2010, 650). Prayer connects us to something bigger than ourselves. Prayer moderates emotion. Prayer offers social support ("knee-mail"). Prayer highlights the importance of the other. Prayer changes our intentions: it primes "caretaking" and motivates us to act benevolently.

Fincham's bottom line: "Guided prayer can be powerful," especially when focused on affirmation, when practiced regularly, and when used both "in conflict AND when things are going well."

## DEEP TRUTHS—AND HALF-TRUTHS—
## ABOUT HUMAN NATURE

### Half Truth: Attitudes Influence Behavior ≃
### Faith Inspires Action

Until about a half century ago, there was a happy concord between the social psychological and Christian perspectives on the importance of our inner attitudes and of the transformative power of a change of heart. My subdiscipline, social psychology, offers other substantial findings that have seemed in close harmony with scripture. Scientific studies of persuasion dating back to World War II have explored how what's inside us sways our actions—how character prompts conduct, how words induce deeds, how opinions inform voting. And if our inner attitudes—our beliefs and feelings about something—determine our outer behaviors, then, to change people's actions, we will want to change their hearts and minds. That's the idea underlying so much of our preaching, teaching, and counseling.

History confirms that *attitudes influence behavior.* Morality fosters virtue. Prejudice produces discrimination. Hate spawns terrorism. Social psychologists have, moreover, shown that our attitudes are especially influential when we are reminded of them—when, before we act, something causes us to stop and remember who we are. Attitudes are also potent when external influences are minimal, and when the attitude is unambiguously specific to the situation.

In such situations, persuasion works. In one experiment, some Caucasian people were exposed to a persuasive message that tanning increases skin cancer risk. A month later, 72 percent of the participants (but only 16 percent of those in a control condition) had lighter skin (McClendon and Prentice-Dunn 2001). Persuasion changed a specific attitude, which changed behavior.

Such findings correlate with Christian axioms: "For it is out of the abundance of the heart that the mouth speaks" (Luke 6:45 NRSV). Blessed are the meek, the merciful, the peacemakers, the pure of heart.

The heart—and the mind—matter. We are to be "transformed by the renewing" of our minds (Romans 12:2). So it was for Paul after his Damascus Road conversion. Elijah, Ezekiel, Isaiah, and Jeremiah likewise experienced an inner transformation. For each of them, a new spiritual consciousness produced new behaviors.

### *Complementary Truth: Attitudes Follow Behavior* ≃ *Actions Feed Faith*

The conclusion seemed clear. Our inner attitudes influence our external behaviors. Then came a new wave of research in social psychology showing that the *opposite* was also true: *our actions influence our attitudes.*

Indeed, we are about as likely to act ourselves into a way of thinking as to think ourselves into action. We not only stand up for what we believe, we believe in what we have stood up for. *Our attitudes follow our behavior.* Such self-persuasion enables religious believers, political advocates, and even future terrorists to believe more strongly in that for which they have witnessed or suffered.

As I explain in *Psychology* (12th edition):

> To get people to agree to something big, start small and build. A trivial act makes the next act easier. Succumb to a temptation and you will find the next temptation harder to resist. In dozens of experiments, researchers have coaxed people into acting against their attitudes or violating their moral standards, with the same result: Doing becomes believing. After giving in to a request to harm an innocent victim—by making nasty comments or delivering presumed electric shocks—people begin to disparage their victim. After speaking or writing on behalf of a position they have qualms about, they begin to believe their own words. (Myers and DeWall 2018, 481)

But this psychological truth also has its counterpart in the language of faith, for faith *is also a consequence of action.* Throughout the Bible, faith is nurtured by obedient action. The Hebrew word for "know" is typically an action verb. To know love, for example, we must not only know about love, we must act lovingly. "Those who do what is true come to the light," said Jesus (John 3:21 NRSV).

Theologians have noted how faith grows as people act on what little faith they have.

- "Faith . . . is born of obedience," said John Calvin (1960, 1.6.2).
- "The proof of Christianity really consists in 'following,'" wrote Søren Kierkegaard (1944, 88).
- "Only the doer of the word is its real hearer," wrote Karl Barth (1956, 1.2, 792)
- To attain faith, argued Pascal ([1670] 1958, 233), "follow the way by which [the committed] began; by acting as if they believed, taking the holy water, having masses said, etc. Even this will naturally make you believe."
- C. S. Lewis agreed: "No conviction, religious or irreligious, will, of itself, end once and for all [these doubts] in the soul. Only the practice of Faith resulting in the habit of Faith will gradually do that" (C. S. Lewis 1981, 61).

Rather than insist that people believe before they pray, Talmudic scholars would urge rabbis to pray and their belief would follow.

## Deep Truth

The two conclusions—that attitudes influence behavior *and* that attitudes follow behavior—are both true. But like the Christian assumptions that faith feeds action, and that actions feed faith, they are half-truths. In both realms, the deeper truth—and the authentic convergence of social psychology and faith—lies in the reciprocal influence of the attitudes of the heart and the actions of the person. In *The Cost of Discipleship*, Dietrich Bonhoeffer captured this dialectic: "*Only he who believes is obedient, and only he who is obedient believes*" (Bonhoeffer 1959, 63).

*Half Truth: Self-Serving Bias ≃ The Sin of Pride*

It is commonly believed that most people suffer low self-esteem—the problem that comedian Groucho Marx portrayed when declaring, "I wouldn't want to belong to any club that would have me as a member." But a mountain of psychological research reveals that most people see and present themselves favorably—with what social psychologists call a "self-serving bias." On self-esteem inventories, people tend to score at the high end of possible scores. Moreover, on subjective and socially desirable dimensions, most see themselves as relatively superior.

This "better than average" phenomenon appears in countless surveys. Most people see themselves as less prejudiced, more ethical, more intelligent, and healthier than their average neighbor or peer. In several surveys, 90 percent or more of drivers rated themselves as better than the average driver. And in "ability to get along with others" nearly all American high school seniors believed themselves above average, with 60 percent rating themselves among the top 10 percent. "How do I love me? Let me count the ways."

Self-serving bias also appears in people's explanations for good and bad outcomes. Most folks readily accept responsibility for their good deeds and successes, while attributing bad deeds and failures to factors beyond their control. This phenomenon of self-serving "attributions" has been evident in laboratory experiments (after people experience success or failure). And it appears in everyday life, as drivers explain accidents, as married partners explain conflicts, as athletes explain victory or defeat, and as students explain good or bad exam grades.

Self-serving bias appears not only in better-than-average self-perceptions and self-serving attributions but also in other well-documented ways. People tend to self-justify their past actions. They display an inflated confidence (not humility) in their judgments. They more readily accept flattering than unflattering descriptions of themselves. And they project their long-term future with Pollyannaish optimism—believing themselves less likely than their peers to suffer cancer, car accidents, or just about any adversity.

Self-disparagement also happens, especially when people are feeling depressed. But on balance, self-serving bias is far more common than low self-esteem. What is more, the human tendency to feel relatively superior can be socially toxic, as people claim greater-than-average credit when their team or work group does well and less-than-average blame when it does not. For the 90-plus percent of college faculty who have thought themselves superior to their average colleague, there surely is a risk of envy and disharmony when merit raises are announced and half receive an average raise or less. Racism and nationalism are collective forms of self-serving bias, as one population views itself as more capable, moral, or deserving than another. The Nazi atrocities were rooted in Aryan pride.

Samuel Johnson recognized the peril of self-serving bias: "He that overvalues himself will undervalue others, and he that undervalues others will oppress them" (S. Johnson 1825, Sermon VI, 344). Dale Carnegie had the same idea: "Each nation feels superior to other nations. That breeds patriotism—and wars" (Carnegie 1993, 122).

For Christians, the newly demonstrated powers and perils of self-serving bias confirm ancient biblical wisdom: "Pride goes before destruction, and a haughty spirit before a fall" (Proverbs 16:18 NRSV). Indeed, theologians have long argued that pride is the original sin, the fundamental sin, the deadliest and most foundational of the seven deadly sins.

The pervasiveness of pride appears in other scriptures: "Who can detect their errors?" asks the psalmist (19:12 NRSV). The Pharisee could thank God that "I am not like other people" (much as we can thank God that we are not like the Pharisee). And St. Paul admonished us to reverse our self-serving pride—to "in humility regard others as better than yourselves" (Philippians 2:3 NRSV).

*Complementary Truth: Self-Esteem Pays Dividends ≍*
*Grace Enables Flourishing*

Social psychology and biblical wisdom are in harmony: self-serving pride is commonplace and perilous. But again, we see that a familiar truth is only half the truth, for the new field of "positive psychology" also documents the benefits of a secure self-esteem and the power of positive thinking. People who express positive self-esteem and self-worth tend to be freer of depression, less prone to substance abuse, less conforming, and higher achieving. Deflate self-esteem, as has been done in experiments (by temporarily being told that others judged them negatively, or they failed on a task), and people will express heightened hostility, including prejudice. Other studies increase people's perceived self-control, as when at-risk youth are given a "growth mindset"—encouraging them to see self-control as a "muscle" that strengthens with exercise. The typical result is increased achievement.

Other studies of "self-efficacy," "intrinsic motivation," and hopeful optimism confirm the benefits of a positive self-view. Believe that your future is beyond your control, and likely it will be. Believe that your initiative can make a difference, and maybe it will. Although there are limits to where self-affirming thinking can take us, self-confidence is adaptive. The Christian counterpart to this further truth is its good news message: compared to enduring others' vacillating opinions about us, the experience of grace—God's acceptance—is a more durable basis for self-acceptance. If a therapist's nonjudgmental acceptance ("unconditional positive regard") is healing, then so much more is God's. If this universe's ultimate transcendent power loves and accepts us, just as we are, then we can be freed from defining our self-worth in terms of our attainments, our prestige, our wealth, and our physical attributes. As self-doubting Pinocchio said to his maker, Geppetto, "Papa, I am not sure who I am. But if I'm all right with you, then I guess I'm all right with me."

Pinocchio's theology was an echo of St. Paul, who, giving up his pretensions, realized, "I no longer have a righteousness of my own, the kind that is gained by obeying the Law. I now have the righteousness that is given through faith in Christ" (Philippians 3:9 GNT). There is wisdom in this tenet of evangelical theology, "with its conviction of sin, its self-despair, and its abandonment of salvation by works," wrote William

James (1890, 168). "To give up one's pretensions is as blessed a relief as to get them gratified."

## Deep Truth

So, yes, the whole truth is that self-serving pride is real and perilous, and also that self-acceptance matters. We experience such acceptance in a supportive friendship or an intimate marriage, in which, even after knowing our foibles, someone accepts us as we are. In such a relationship, whether with a human or with God, we no longer need to explain and defend ourselves, but simply to accept our acceptance. As the psalmist experienced, "LORD, I have given up my pride and turned away from arrogance. . . . I am content and at peace" (Psalm 131:1 GNT).

## Half Truth: Human Cognitive Powers ≃ Humans in the Image of God

From womb to tomb, humans display remarkable analytic powers. Thanks to our head's three pounds of unfathomably complex neural connections, we display astounding abilities beginning before birth. Newborns already display recognition of their mother's voice, and soon can discriminate her smell. From birth on, we absorb and process information and rapidly begin to form concepts, retain memories, and creatively solve problems.

Consider our human capacity for language. Well before being able to ride a bike, preschoolers are soaking up language with a skill that would put a college foreign language student to shame. By age two and beyond, they are creating grammatically intelligible sentences and comprehending the more complex sentences spoken to them. Or consider your own capacity for conversation—during which you manage, simultaneously and spontaneously, to choose and order words; monitor your muscles, tone, and gestures; and then shoot pulsating air molecules through space—enabling you to wirelessly transmit ideas from one head (yours) to another. Those pulsating air-waves contain information!

We are, indeed, *Homo sapiens*, the wise species. We are, as Hamlet, exclaimed, "infinite in faculties! . . . How like a god!"

Much earlier, the psalmist (8:5) was similarly awestruck. Humans, he rhapsodized, are but "a little lower than God." Pascal's *Pensees* also emphasized the cognitive powers of our species: "Man's greatness lies in his power of thought" (Pascal [1670] 1958, 346; literal translation, "Thought constitutes the greatness of man").

As theologians have stressed, we are made in the image of God. We are the summit of God's creative work. We are God's own children.

### Complementary Truth: Illusory Thinking ≃ The Finite Creature

So, there is a harmony between the witness of science and of scripture: the human creature is worthy of awe. And now for the rest of the story: to err is human.

Thanks to the new science of "cognitive social psychology," we have amassed evidence of predictable pitfalls of human thinking—reasons for unreason. Humans have an enormous capacity for automatic, efficient thinking. But our cognitive efficiency, though generally adaptive, comes at a price—rather like the confusing visual illusions that are by-products of our efficient visual system. Some examples:

**Overconfidence.** We often overestimate the accuracy of our judgments. We much more easily imagine why we might be right than wrong.

**Confirmation bias.** We more often search for information that can confirm our beliefs than for information that can disconfirm them.

**Availability heuristic.** When given compelling anecdotes or even useless information, we often ignore useful statistical information. This is partly due to the later ease of recall of vivid (and cognitively available) information.

**Illusions of correlation and personal control.** It is tempting to perceive correlations where none exist (*illusory correlation*) and to think we can predict or control chance events (the *illusion of control*).

**Moods infuse judgments.** Our moods trigger memories of experiences associated with those moods. Good and bad moods also color our interpretations of current experiences.

**Flawed self-knowledge.** We often do not know why we do what we do. We also may deny our attitudes have changed when they have, deny real influences upon us, and fail to predict our future behavior.

**Preconceptions sway interpretations.** We often see and later recall what we expect.

**Self-fulfilling beliefs.** Our beliefs and expectations may generate their own reality. Thus, in everyday life, we often get what we expect. If we expect someone to be hostile, our behavior may trigger unfriendliness.

In such ways, we sometimes seem, as T. S. Eliot lamented, headpieces "filled with straw."

And that would come as no surprise to theologians who have long assumed that we are finite creatures of the one who declares, "I am God, and there is none like me" (Isaiah 46:9 KJV). Always, we see reality in a mirror, dimly. "For as the heavens are higher than the earth, so are my ways higher than your ways and my thoughts than your thoughts" (Isaiah 55:9 KJV).

### Deep Truths

In these and other ways, we see that in both psychological science and Christian belief the whole truth seems to lie in complementary propositions. Faith is a source *and* a consequence of action. Self-serving pride is basic to our nature, *yet* self-acceptance borne of grace enables human flourishing. We are awesome creatures made in God's image, *yet* also finite and error-prone.

Faith always seeks understanding in the language of the day. Today, that includes the revelations of psychological science, which can help us remember various dual lessons of scripture and theology. To ask whether faith or action comes first (they are each chicken and egg), whether pride or self-disparagement is the problem, and whether we are wise or foolish, is akin to asking which blade of a pair of scissors is more essential.

To paraphrase what I have previously written:

> Faced with these pairs of complementary ideas, framed either psychologically or theologically, we are like someone stranded in a deep well with two ropes dangling down. If we grab either one alone we sink deeper into the well. Only when we hold both ropes can we climb out, because at the top, beyond where we can see, they come together around a pulley. Grabbing only the rope of attitudes → behavior or behavior → attitudes, of self-serving pride or healthy self-esteem, of human rationality or irrationality, plunges us to the bottom of the well. So we grab both ropes, perhaps without fully grasping how they come together. In doing so, we may be comforted that in both science and religion, accepting complementary principles is sometimes more honest than an oversimplified concordism that ignores half the evidence. For the scissors of truth, we need both blades. (Myers 2010a, 62)

# The Way Ahead for Psychological Science and Christian Faith

A s ENTHUSIASTS for psychological science, we are aware of psychology's past achievements and its potential for further exciting developments. We are also serious Christians. In this volume, we have tried to present evidence that Christian faith and psychological science can share mutual insights and enrichments on a range of topics. This "insights-and-enrichments" approach may be seen as related to, and as a further development of, some past suggestions variously described as *perspectivalism*, *complementary perspectives* (or simply *complementarity*), and the *levels-of-explanation* view. Of these past approaches, complementarity comes closest to our present proposal. Before turning to complementarity, we review what we see as some problems with other past and present approaches to the relationship between science and faith.

## SCIENCE-FAITH CONFLICT IS NOT INEVITABLE

We have argued that some "knee-jerk reactions" from Christians in the past, seeing conflicts between each new development in science and their cherished Christian beliefs, have been unnecessary and counterproductive. We provided some illustrations of these false conflicts, which Pablo de Felipe and one of us (MAJ) unearthed while gathering materials for other purposes (de Felipe and Jeeves 2017). For example, we believe that no lasting benefit for Christian faith was gained by church theologians who claimed that scripture contradicted the growing scientific consensus about the spherical nature of the earth. This conflict was not necessary, because the shape of the earth was essentially a scientific question rather than a theological question. From these instances, lessons emerged that we suggested were worth keeping in mind to help avoid similar errors today, especially when asking questions about human nature.

At a philosophical level, we hold that conflicts between science and faith are not inevitable, because ultimately all truth is God's truth. We have held this position for a long time, as this quotation demonstrates:

> As a Christian, it is my belief that what God has chosen to give to man down the ages through his special messengers will not ultimately conflict with what he has encouraged us to discover as we exercise his gifts of mind and hand in exploring the created order, including, of course, man himself, his experience and his behaviours. (Jeeves 1976, 18)

## SCIENCE-FAITH CONCORDISMS ARE APPEALING BUT DANGEROUS

Conflict is only one of the recurrent themes that historians have identified in describing the relationship between science and religion. Another, though disguised at times in various garbs, has been the desire to produce concordisms—harmonizing the current scientific claims with the current interpretation of scripture—to use the prestige of science to bolster Christian beliefs. In the chapter on the conflict motif in the history of science and faith relations, we argued that, although the popularizers of the conflict model have habitually portrayed Christians as conspiring against science, the truth was frequently the opposite. The prestige of the natural philosophers, and then of the scientists—today's natural philosophers—has been so high that religious leaders have sought to mobilize their support, at times at the price of making intellectual concessions to achieve concordism. Using historical examples, we examined whether the carefully constructed concordisms of one age sowed the seeds of later conflicts.

To illustrate the unforeseen consequences of this approach, we referred to the subtle concordism of dualistic anthropology. Body-soul dualism was embraced by church leaders in the second century as a self-evident "scientific" truth from natural philosophy. Once dualism was incorporated deeply into Christian theology, this concordism persisted without serious challenge until the twentieth century. With the rise of medical physiology and then neuroscience, the vast majority of contemporary scientists have discarded dualism and adopted a monist "physicalism," leaving dualistic theologians clinging to an outdated science. Thus, *a new conflict has emerged from the ashes of the previous concordism.*

But once again, this conflict is not really necessary. Over the past century, biblical scholars also began to move away from a dualistic anthropology in order to recover a more "holistic" Hebrew view of the human person. The rejection of Platonic dualism provides an opportunity for theologians and psychologists to work together in engaging the neuroscience findings that support a fundamental mind-brain and mind-body unity of the human person. In our view, the most helpful way to move forward is to recognize the mysterious duality of our mental life and physical body, while accepting our *essential psychobiological unity as whole, complete persons.* Because both aspects of the human person are important and relevant, we use the rather clumsy label *dual-aspect monism* to characterize our view (See MacKay 1991; Jeeves [1997] 2006; Benovsky 2016).

We also described the problems that a second major example of concordism produced in psychology's short history. In the early 1800s, when *phrenology* was the "new science" of the mind, several Christian leaders built a theological system based on phrenology. When the scientific basis of phrenology was discredited, that theological system collapsed. We also pointed to the current interest in neurotheology, which in some ways resembles the enthusiasm for phrenology among some earlier Christians. In our view, concordisms built upon the notion that humans are "hardwired for religion"—or the presumed existence of a "God spot" in the brain—will suffer the same fate as those built on phrenology.

## Advances in Both Science and Theology Destroy Concordisms

The above examples show that concordisms are undermined by changing views among scientists. Concordisms are also threatened by changing views among theologians, as John Walton, Old Testament professor at Wheaton College, warned:

> New insights and new information can emerge at any time. Several hundred years ago, renewed access to the original languages had significant impact on biblical interpretation. In recent decades, the availability of documents from the ancient world has provided a remarkable resource for our reading of the biblical text. We dare not neglect these tools when they can contribute so significantly to our interpretation. (Walton 2015, 12)

The impact of developments in biblical scholarship is well illustrated in a recent volume bringing together a range of views held by Christians about issues such as the age of the earth and the evolutionary accounts of the emergence of humans. It includes a contribution by a trio of leading Old Testament scholars highlighting a point we have made repeatedly in this book, namely, that we must pay attention not only to advances in science but also to advances in biblical scholarship. This essay refers specifically to the hazards of *premature concordisms*, and echoes our criticism of the loose use of words like "concordism" and "integration":

> We now need to examine this hermeneutical principle of extended authority as it pertains to scientific readings in the Bible. One of the common approaches to Scripture that

attempts to extend meaning beyond the biblical author's original intentions is called *concordism*. Concordist interpreters claim there is a convergence between God's Word and God's world and suggest ways that a more sophisticated scientific understanding of the world can be integrated with statements of Scripture—admittedly applying meaning to the words of Scripture that the author would never have been aware of. Such extended meanings can claim no authority since they do not derive from inspired sources. *They cannot justifiably represent claims to perceive meanings that God intended, because they are not meanings that are independent of our own imagination.* (Lemke, Walton, and Samples 2017, 31–32, emphasis added)

## Is Science–Faith Integration a Worthy Goal?

If we are to move beyond conflicts and concordisms, what direction should we follow? One popular approach is to attempt the *integration* of psychology and Christianity. This approach received considerable publicity forty years ago when philosopher Stephen Evans helpfully reviewed the various ways that different psychologists conceptualized the relationship between their science and their faith (Evans 1977). A pervasive theme throughout his writings was the idea of "integrating" psychology and Christian belief. On the surface, this sounds like an appropriate position to be taken by Christians who believe that the God who "upholds the universe by the word of his power" (Hebrews 1:3 ESV) is also the God whose world we are investigating as scientists.

At the time that Evans was writing, few of the developments described in this book within neuropsychology, cognitive science, and evolutionary psychology had taken place. The focus of most of the discussions was personality theory, counseling, and perhaps social psychology. At that time, the language used by psychologists working in these specialties did not sound all that different from the language used in scripture when describing human behavior. Perhaps it was understandable to look for integration between the two domains. However, the technical language used by some contemporary psychologists looks almost as different from the language of scripture as does the language in nuclear physics and evolutionary biology. Therefore, the temptation to attempt an integration of the two is more easily resisted.

Of course, it is not only psychology that has changed in the past four decades. Significant advances have been and are being made in biblical scholarship and in theology. Those who tried to demonstrate integration between psychology and Christian belief have frequently used outdated psychology or outdated biblical scholarship, or both. In those cases, any integration attempt is on shaky ground and will soon need to be discarded.

We need to remember that today's science is just that: *today's* science. By its nature, the current scientific consensus is destined to change and develop in the light of new research—a point noted by Nobel laureate Venki Ramakrishnan, the current president of the Royal Society, whose views we quoted earlier. Also, the current theological consensus is likely to change, as biblical scholars reevaluate their interpretations. We believe that it would be unwise to propose any theory of human nature that depends on combining today's psychological science and today's biblical interpretations as if they were set in stone.

## Diverse Views about Integration

The concept of integration was widely used in discussions of psychology and Christianity for several decades in the latter half of the twentieth century. Our review of the uses of the term, as well as actual attempts at integration, led us to conclude that *the language of integration has served its purpose and outlived its usefulness.* Moving forward, we believe that the term should be used sparingly or not at all.

As one indication of the problems with the language of integration, consider the articles published in the recent special issues of the *Journal of Psychology and Theology* and the *Journal of Psychology and Christianity* that were devoted to integration. The diversity of views about integration expressed in those articles was so great—ranging from authors advocating the total intermixing of psychology and theology to authors arguing that integration attempts are misguided—that it is difficult to see a common theme on which to build future progress. Across these articles, we detected three main ways in which the language of integration was used:

- *Complete integration* as an unrestricted intermixing of psychology and Christianity, producing an amalgam that is labeled "Christian psychology."
- *Selective integration* as an incorporation of selected aspects of psychology into Christian thought, excluding topics that don't fit well with Christianity.
- *Avoiding integration.* This view was represented by Mary Stuart van Leeuwen, who wrote, "My biggest challenge has been trying to get rid of the word integration. I don't like it because I prefer to talk about the unity of faith and learning" (Lewis Hall et al. 2006, 332).

However, we should note that Eric Johnson, who has referred to integration as "in principle, a calling from God," describes these "three faces of integration" using different terms: *maximal integration, strategic integration*, and *dissociative integration* (E. Johnson 2011). As Everett Worthington and Todd Hall wrote in the introduction to one of the special issues, "The world has changed. Integration also has changed. In fact, integration is much more diverse now than the virtual mono-approach of 40 years ago" (Worthington and Hall 2012). Because *integration* means so many different things to different people, we believe it is time to move beyond the language of integration, shifting to different terminology that more precisely describes one's view of the relationship between psychology and faith.

## SCIENCE AND FAITH AS NONOVERLAPPING MAGISTERIA?

For Christians who have grown disillusioned with all three of the models we have discussed—conflict, concordisms, and integration—what alternatives exist? How should we properly relate statements about the same set of events, when those statements originate from within the two separate and distinct language domains of science and faith? As we noted earlier, the founding father of big bang cosmology, Georges Lemaître, thought the solution was simple. Because statements from science and from theology come from within two different areas of thought, based on different assumptions and serving different purposes, Lemaître saw no conflict between his faith and his scientific work (Aikman 1933, 18). Therefore, he had no need for concordisms or attempts at integration. Lemaître's views were expanded and popularized by Stephen Jay Gould in *Rocks of Ages: Science and Religion in the Fullness of Life* (Gould 1999). Gould argued that science and religion represent "nonoverlapping magisteria"—separate domains of authority that can coexist peacefully. Religion is concerned with our moral world, while science investigates the natural world. Ideally, each would be content to stay within its separate sphere of influence, treating the other with polite indifference and respectful noninterference.

This approach, also known as *compartmentalism*, or in Ian Barbour's (2000) typology as *independence*, is a view held by some scientists who are Christians. It allows them to be hard-nosed reductionists when engaged in their roles as scientific investigators, while still leaving room for religious beliefs and practices in their life outside the laboratory. But in a way, this approach is deeply unsatisfying. If there is a God who is the source of all things and the ruler of the universe, it is difficult to see how the study of this God—theology—can be excluded from the domain of the natural world. In fact, the two magisteria overlap, in the sense that they both have interests in and explanations for some of the most important issues for humans today.

## COMPLEMENTARITY AS AN ALTERNATIVE TO
## NONOVERLAPPING MAGISTERIA

In our reading of the history of science-faith relations, the conflict model and the attempts at concordism have both seemed to do more harm than good. In our view, the jury is still out on the value of the integration model. Is there a better way of thinking about how properly to seek to relate the knowledge gained from psychological science and that which comes from our Christian faith? We noted that Stephen Evans, toward the end of his book *Preserving the Person*, had a section headed "A Model for Integration" (Evans 1977). In this section, he summed up his own position on the relation between science and faith. He asserts that the view one group of authors (e.g., Jeeves, MacKay, and Myers) have put forward since the 1960s and 1970s—which Evans characterizes as the *Perspectivalist/Limiter of Science* view—offers rich possibilities for "integrating contemporary scientific perspectives with the traditional personalistic account." He continued:

> It seems to me therefore that the Perspectivalist/Limiter of Science position offers great promise as a working model for reconciling contemporary sciences of man with biblical personalism. Such a view allows for a unified view of the person which is congenial to the Bible's emphasis on the unity of the person and the resurrection of the body. (Evans 1977, 154)

This "unified, dual-aspect" view has attracted us for a long time, as is evident from this quotation:

> We have seen that different types of fact demand different types of treatments; that, in particular, scientific thinking and faith are distinguished partly on this account. Our whole approach in the two cases, it has been suggested, must be different, requiring different attitudes of mind. (Jeeves 1953, 21)

The *perspectivalist* approach reflected in that quotation is sometimes labeled *complementarity*—referring to two complementary accounts of the same events given from different perspectives—or *levels of explanation*. As David Myers described it four decades ago,

> If we are to understand why scientific and religious explanations need not be considered mutually exclusive, we must recognize that there are a variety of possible ways by which one might explain a given event, all of which can be simultaneously true in their own terms. An exhaustive description at one level does not invalidate explanation at some other level. (Myers 1978, 11)

This truth is driven home by Donald MacKay's famous example of an electric sign board that displays a flashing advertisement (MacKay 1974, 36–38). A technician could provide a description of the board in terms of the electrical components—a description so detailed that someone else could reproduce the sign at a different location. But that description would be incomplete, because it ignores the message of the sign and the purpose for which the sign was constructed. Likewise, a description of the sign in terms of its message would be accurate and complete at one level but would ignore the inner workings of the physical components. In MacKay's words, "The two are not explanations in the same sense. They are answers to different questions, and both may, in fact, be entirely valid" (MacKay 1960, 10). David Myers summed up the point this way:

> So we see that in Christian belief, much as in contemporary psychology, the whole truth seems best approximated by *complementary propositions*. (Myers 1987, 139, emphasis added)

One of the great values that we see in this approach is that it removes much of the anxiety about science that many Christians feel—especially

Christians who subscribe to a "god-of-the-gaps" viewpoint—as science seems to answer more and more of the questions once thought to be in the domain of theology. As Donald MacKay explained:

> But if once we recognise that at least most theological categories are not "in the same plane" (in the same logical subspace) as most scientific categories, there is no longer any theological merit hunting for gaps in the scientific pattern. Gaps there are in plenty. But . . . it would seem to be the Christian's duty to allow—indeed to help—these gaps to fill or widen as they will, in humble and cheerful obedience to the truth as God reveals it through our scientific discipline, believing that to have theological stakes in scientific answers to scientific questions is to err in company with those unbelievers who do the like. (MacKay 1953, 163)

## Complementarity Requires Rigorous Thinking

While advocating the importance and helpfulness of the notion of *complementarity* (here and in chapter 5), we believe that it is still necessary to remember the cautionary note that one of us (MAJ) wrote on this topic twenty years ago:

> This principle of *complementarity*, first enunciated for physics by Niels Bohr, is an analogy, not in any sense a proof of the necessity of complementary Christian and scientific viewpoints. However satisfied we may be that the two pictures are *compatible*, only the facts of experience can convince us that both are *necessary*. We are dealing here with a logical point, not a scientific one, but it is one which is open to easy abuse and misunderstanding. For this reason we need to see clearly the conditions under which it can legitimately be used. Unless

we do this, it could easily become an escape hatch which we use when we get into a tight corner in discussions concerning the relation of science and faith. (Jeeves and Berry 1998, 81, emphasis added)

As Donald MacKay warned:

Complementarity is no universal panacea, and it is a relationship that can be predicated of two descriptions only with careful safeguards against admitting nonsense. Indeed the difficult task is not to establish the possibility that two statements are logically complementary, but to find a rigorous way of detecting when they are not. (MacKay 1953, 163)

What are the conditions under which two or more descriptions may be considered logically complementary? According to MacKay, the descriptions must meet these conditions:

(1) that they purport to have a common reference; (2) that each is in principle exhaustive (in the sense that none of the entities or events comprising the common reference need to be left unaccounted for); yet (3) they make different assertions because (4) the logical preconditions of definition and/or of the use (that is the context in which they are set) of concepts or relationships in each are mutually exclusive, so that the significant aspects referred to in one are necessarily omitted from the other. (Barbour 1974, 77–78)

Concluding our statement of caution, we pointed out, "Before religious and scientific statements could be debated as rivals, it would be necessary to establish that they are not in fact complements. It is also of course equally necessary to realise and to recognise that proof of complementarity would not establish that either account is true" (Jeeves and Berry 1998, 81).

## SUPPORT FOR THE COMPLEMENTARITY APPROACH: INSIGHTS AND ENRICHMENTS

The *perspectivalist* position has gained support from recent analyses by writers such as Fraser Watts, Everett Worthington, and David Myers, who, in his chapter in this volume, wrote:

> Many of us Christians in psychology seek to explore God's human creation with a spirit of curiosity and humility. . . . And as we pursue an authentic convergence of scientific and religious wisdom, we can tolerate seeming conflicts, without resorting to oversimplified concordisms.

In the second half of this book, we took up the theme of insights and enrichments. We noted that Everett Worthington had already suggested a number of examples of enrichments (Worthington 2010), and Fraser Watts contributed other examples, reminding us that the process of mutual insights and enrichments is a two-way street (Watts 2012). For our part, with our specialist knowledge in neuropsychology, cognitive science, and evolutionary psychology, we noted, by way of illustration, a number of possible insights and enrichments from these areas of science that are relevant not just for Christian doctrine but also practice in the Christian life.

One specific topic that we chose to use as an example was new insights into spirituality. Developments in neuropsychology have emphasized the psychobiological unity of the human person, providing fresh insights into the connection between spiritual experiences and the intact functioning of the brain. Some examples:

- ▶ Recent research into the changes in felt and lived spirituality among patients with progressive Alzheimer's disease
- ▶ Clear links between some peak religious experiences and unusual activity in the temporal lobes of the brain
- ▶ Reduced capacity for moral reasoning and moral behavior following damage to the frontal lobes of the brain
- ▶ Reduced capacity for interpersonal trust in conditions such as Capgras syndrome, suggesting that trust is embodied in brain activity
- ▶ Recent research into possible changes in spirituality in persons with Parkinson's disease

It is abundantly clear that our spirituality is firmly embodied in our psychobiological makeup, and that has profound implications for the use of body-soul dualistic language when referring to spiritual experiences. The insights from this kind of research in turn reminded us of the constant need for patience and compassion in all our relationships as members of one body, the body of Christian fellowship that is the church on earth. A related example concerns the nature of the church and the importance of taking full account of the new insights afforded by advances in neuropsychology for worship and community life. Here we made detailed references to the work of Warren Brown and Brad Strawn in their book *The Physical Nature of Christian Life* (Brown and Strawn 2012).

When we turned to evolutionary psychology, we noted how it could shed new light on a topic discussed over many centuries by theologians and philosophers as well as scientists, namely, what makes humans distinctive from other animals. We saw that, for centuries, the traditional way of defining "humanness" was to seek particular criteria allegedly unique to the human condition: bipedalism, opposable thumbs, toolmaking, learning ability, abstract thought, shame, play, or artistic sense. We noted how so many of these supposed criteria were no longer considered unique to humans. Despite the concern that some Christians have about an apparent narrowing of the gap between us and some of our nonhuman primate relatives, there are in fact no great theological issues at stake in this research. A Christian can be enthusiastically open-minded about developments in evolutionary psychology, seeing in some of its discoveries new insights into what does and does not make us uniquely human.

While evolutionary biology and evolutionary psychology contend that we are descended from primate ancestors, nevertheless, scripture affirms that we differ substantively from them by virtue of being "in God's image." So, what does this mean? Having participated in a meeting with geneticists, evolutionary psychologists, cognitive scientists, neuropsychologists, philosophers, and theologians, one of us (MAJ) wrote, in the final chapter of a book summing up their diverse views, and commenting on the contribution of theologian Anthony Thiselton:

> After his very detailed exposition of what is meant by claims that have been made that humans are made in the image and likeness of God, an exposition that covers reflections over the past two millennia and before, as well as more recent trends, Thiselton looks again at the three key aspects of what it means to be made in the image of God—relationality, representation, and vocation, or attainment—in the light of some of the contributions from the scientists in earlier chapters. (Jeeves 2015a, 237)

What is noteworthy here is that there is no reference to basing human distinctiveness on special capacities that humans possess and other animals do not. Thiselton had heard sufficient evidence from the scientists to realize that that was not a helpful way to go.

We should marvel that the Creator of the whole of creation has, in his wisdom, created a world, including ourselves, where in his good time we came to possess remarkable abilities for a personal relationship with our Creator. The challenge remains: Do we capitalize on these abilities and enter into this personal relationship that our Creator—who, we discover, is also our Redeemer—offers us?

# References

Adams, Jay E. 1970. *Competent to Counsel: Introduction to Nouthetic Counseling*. Grand Rapids: Zondervan.

Aikman, Duncan. 1933. "Lemaître Follows Two Paths to Truth." *New York Times*, February 19.

Albright, Carol R., and James N. Ashbrook. 2001. *Where God Lives in the Human Brain*. Naperville, IL: Sourcebooks.

Alexander, Denis. 2017. "Adam and the Genome: Some Thoughts from Denis Alexander." February 2. biologos.org.

Allman, J. M., N. A. Tetreault, A. Y. Hakeem, K. F. Manaye, K. Semendeferi, J. M. Erwin, S. Park, V. Goubert, and P. R. Hof. 2010. "The von Economo Neurons in Frontoinsular and Anterior Cingulate Cortex in Great Apes and Humans." *Brain Structure and Function* 214 (5–6): 495–517.

Allport, Gordon W. 1950. *The Individual and His Religion: A Psychological Interpretation*. New York: Macmillan.

Argyle, Michael. 1958. *Religious Behavior*. London: Routledge.

———. 1964. "Seven Psychological Roots of Religion." *Theology* 67 (530): 333–39.

Arnold, Bill T. 2004. "Soul Searching Questions about 1 Samuel 28: Samuel's Appearance at Endor and Christian Anthropology." In *What about the Soul? Neuroscience and Christian Anthropology*, edited by J. B. Green, 75–83. Nashville: Abingdon Press.

Augustine. 1954. *Saint Augustine: The City of God*. Translated by Gerald G. Walsh and Daniel J. Honan. Washington, DC: Catholic University of America Press.

Babbage, Charles. 1837. *The Ninth Bridgewater Treatise, a Fragment*. London: John Murray.

Baddeley, A. 1992. "Working Memory." *Science* 255 (5044): 556–59.

Barbour, Ian G. 1974. *Myths, Models, and Paradigms: The Nature of Scientific and Religious Language*. London: SCM Press.

———. 2000. *When Science Meets Religion: Enemies, Strangers, or Partners?* San Francisco: HarperCollins.

Barlow, Nora, ed. 1967. *Darwin and Henslow: The Growth of an Idea*. London: Bentham-Moxon Trust, John Murray.

Barnett, Keri L., Rodney L. Bassett, Julia P. Grimm, and Cheryl L. Repass. 2012. "Revisiting the Issue of Influential Sources in the Integration of Psychology and Theology: Are We There Yet?" *Journal of Psychology and Theology* 40 (2): 16–20.

Barr, James. 1993. *Biblical Faith and Natural Theology: The Gifford Lectures for 1991, Delivered at the University of Edinburgh*. Oxford: Clarendon Press.

Barrett, Justin. 2011. *Cognitive Science, Religion, and Theology: From Human Minds to Divine Minds*. West Conshohocken, PA: Templeton Press.

Barrett, Justin L., and Tyler S. Greenway. 2017. "*Imago Dei* and Animal Domestication." In *Human Origins and the Image of God: Essays in Honor of J. Wentzel van Huysteen*, edited by Christopher Lilley and Daniel J. Pedersen, 64–81. Grand Rapids: Eerdmans.

Barrett, Justin, and Matthew Jarvinen. 2015. "Cognitive Evolution, Human Uniqueness, and the *Imago Dei*." In *The Emergence of Personhood: A Quantum Leap?* edited by M. A. Jeeves, 163–83. Grand Rapids: Eerdmans.

Barth, Karl. 1956. *Church Dogmatics*, Vol. I, Part 2, edited by G. W. Bromiley and T. F. Torrance. Translated by G. T. Thomson and H. Knight. Edinburgh: T. & T. Clark.

———. 1960. *Church Dogmatics*, Vol. III, Part 2, edited by G. W. Bromiley and T. F. Torrance. Translated by H. Knight, G. W. Bromiley, J. K. S. Reid, and R. H. Fuller. Edinburgh: T. & T. Clark.

Bartlett, Frederic C. 1932. *Remembering: A Study in Experimental and Social Psychology.* Cambridge: Cambridge University Press.

———. 1950. *Religion as Experience, Belief, Action.* London: Oxford University Press.

Basten, Ulrike, Guido Biele, Hauke R. Heekeren, and Christian J. Fieback. 2010. "How the Brain Integrates Costs and Benefits during Decision Making." *Proceedings of the National Academy of Sciences* 107 (50): 21767–72.

Bate, Sarah, and Jeremy J. Tree. 2017. "The Definition and Diagnosis of Developmental Prosopagnosia." *Quarterly Journal of Experimental Psychology* 70 (2): 193–200.

Bateson, Melissa, Daniel Nettle, and Gilbert Roberts. 2006. "Cues of Being Watched Enhance Cooperation in a Real-World Setting." *Biology Letters* 2 (3): 412–14.

Bauckham, Richard. 2006. *Jesus and the Eyewitnesses: The Gospels as Eyewitness Testimony.* Grand Rapids: Eerdmans.

———. 2015. "The Gospels as Testimony to Jesus Christ: A Contemporary View of Their Historical Value." In *The Oxford Handbook of Christology*, edited by Francesca A. Murphy and Troy A. Stefano, 55–71. Oxford: Oxford University Press.

Beck, James R. 2006. "Integration: The Next 50 Years." *Journal of Psychology and Christianity* 25 (4): 321–30.

Benovsky, Jiri. 2016. "Dual-Aspect Monism." *Philosophical Investigations* 39 (4): 335–52.

Benson, H., J. A. Dusek, J. B. Sherwood, P. Lam, C. F. Bethea, W. Carpenter, S. Levitsky, P. C. Hill, D. W. Clem Jr., M. K. Jain, D. Drumel, S. L. Kopecky, P. S. Mueller, D. Marek, S. Rollins, and P. L. Hibberd. 2006. "Study of the Therapeutic Effects of Intercessory Prayer (STEP) in Cardiac Bypass Patients: A Multicenter Randomized Trial of Uncertainty and Certainty of Receiving Intercessory Prayer." *American Heart Journal* 151 (4): 934–42.

Berger, Klaus. 2003. *Identity and Experience in the New Testament.* Minneapolis: Fortress Press.

Blocher, Henri. 1984. *In the Beginning.* Leicester: InterVarsity Press.

Bloom, Paul. 2004. *Descartes' Baby: How the Science of Child Development Explains What Makes Us Human.* New York: Basic Books.

Boa, Kenneth D. 1994. "Theological and Psychological Accounts of Human Needs: A Comparative Study." PhD dissertation, Oriel College, University of Oxford.

Bonhoeffer, Dietrich. 1959. *The Cost of Discipleship.* New York: Touchstone/ Simon & Schuster.

Boring, Edwin G., Herbert S. Langfeld, and Harry P. Weld. 1948. *Foundations of Psychology.* New York: John Wiley & Sons.

Bouma-Prediger, Steve. 1990. "The Task of Integration: A Modest Proposal." *Journal of Psychology and Theology* 18 (1): 21–31.

Brahe, Tycho. 1913–29. *Tychonis Brahe Dani Opera Omnia.* Edited by J. L. E. Dreyer. Copenhagen: Nielsen and Lyciche.

Brennan, Robert. 2015a. *Describing the Hand of God: Divine Agency and Augustinian Obstacles to the Dialogue between Theology and Science.* Eugene, OR: Wipf and Stock Publishers.

———. 2015b. "Re-examining Tertullian and Augustine's Relationship for the Theology Science Dialogue." *Science and Christian Belief* 27 (1): 84–100.

Broca, Paul. 1861a. *Perte de la Parole, Ramollissement Chronique et Destruction Partielle du Lobe Antérieur Gauche du Cerveau* [Loss of speech, chronic softening, and partial destruction of the anterior left lobe of the brain]. *Bulletin de la Société d'Anthropologie,* 2: 235–38.

———. 1861b. *Sur le Principe des Localisations Cérébrales* [On the principle of cerebral localizations]. *Bulletin de la Société d'Anthropologie* 2: 190–204.

Brooke, John H. 1991. *Science and Religion: Some Historical Perspectives.* Cambridge: Cambridge University Press.

———. 2009. "Genesis and the Scientists: Dissonance among the Harmonizers." In *Reading Genesis after Darwin,* edited by S. C. Barton and D. Wilkinson, 93–109. London: Oxford University Press.

Brown, Laurence B, ed. 1985. *Advances in the Psychology of Religion.* Vol. 11. Elmsford, NY: Pergamon Press.

Brown, Laurence B., and Joseph Paul Forgas. 1980. "The Structure of Religions: A Multi-Dimensional Scaling of Informal Elements." *Journal for the Scientific Study of Religion* 19 (4): 423–31.

Brown, Warren S. 2000. "Wisdom and Human Neurocognitive Systems." In *Understanding Wisdom*, edited by W. S. Brown, 193–214. West Conshohocken, PA: Templeton Foundation Press.

Brown, Warren S., Nancey Murphy, and H. Newton Malony. 1998. *Whatever Happened to the Soul? Scientific and Theological Portraits of Human Nature*. Minneapolis: Fortress Press.

Brown, Warren S., and L. K. Paul. 2000. "Psychosocial Deficits in Agenesis of the Corpus Callosum with Normal Intelligence." *Cognitive Neuropsychiatry* 5: 135–57.

———. 2015. "Brain Connectivity in the Emergence of Capacities of Personhood: Reflections from Callosal Agenesis and Autism." In *The Emergence of Personhood: A Quantum Leap?* edited by M. A. Jeeves, 104–19. Grand Rapids: Eerdmans.

Brown, Warren S., and Brad D. Strawn. 2012. *The Physical Nature of Christian Life: Neuroscience, Psychology, and the Church*. Cambridge: Cambridge University Press.

———. 2017. "Beyond the Isolated Self: Extended Mind and Spirituality." *Theology and Science* 15 (4): 411–23.

Bruce, Charles, Robert Desimone, and Charles G. Gross. 1981. "Visual Properties of Neurons in a Polysensory Area in Superior Temporal Sulcus of the Macaque." *Journal of Neurophysiology* 46 (2): 369.

Bruner, Jerome, and Carol F. Feldman. 1996. "Group Narrative as a Cultural Context of Autobiography." In *Remembering Our Past: Studies in Autobiographical Memory*, edited by D. C. Rubin, 291–317. New York: Cambridge University Press.

Burkhardt, Richard W. 2005. *Patterns of Behavior: Konrad Lorenz, Niko Tinbergen, and the Founding of Ethology*. Chicago: University of Chicago Press.

Burns, Jeffrey, and Russell Swerdlow. 2003. "Right Orbitofrontal Tumor with Pedophilia Symptom and Constructional Apraxia Sign." *Archives of Neurology* 60 (March): 437–40.

Buss, David. 1998. *Evolutionary Psychology: The New Science of Mind*. Boston: Allyn & Bacon.

Butler, Paul M., Patrick McNamara, and Raymon Durso. 2011. "Side of Onset in Parkinson's Disease and Alterations in Religiosity: Novel Behavioral Phenotypes." *Behavioural Neurology* 24 (2): 133–41.

Byrd, Randolph C. 1988. "Positive Therapeutic Effects of Intercessory Prayer in a Coronary Care Unit Population." *Southern Medical Journal* 81 (7): 826–29.

Byrne, Richard W. 2006. "Evolutionary Psychology and Sociobiology: Prospects and Dangers." In *Human Nature*, edited by M. A. Jeeves, 84–105. Edinburgh: Royal Society of Edinburgh.

Byrne, Richard, and Andrew Whiten. 1988. *Machiavellian Intelligence: Social Expertise and the Evolution of Intelligence in Monkeys, Apes, and Humans*. Oxford: Clarendon Press.

Cacioppo, John T., and Gary G. Berntson, ed. 2005. *Social Neuroscience: Key Readings*. New York: Psychology Press.

Calvin, John. 1847. *Commentaries on the First Book of Moses Called Genesis*. Translated by John King. Edinburgh: Calvin Translation Society.

———. 1960. *Institutes of the Christian Religion*, edited by J. T. McNeill. Translated by F. L. Battles. Philadelphia: Westminster Press.

Carnegie, Dale. 1993. *The Leader in You: How to Win Friends, Influence People, and Succeed in a Changing World*. New York: Pocket Books.

Carson, Donald. 1994. "When Is Spirituality Spiritual? Reflections on Some Problems of Definition." *Journal of the Evangelical Theological Society* 37 (3): 381–94.

Carter, John D. 1977. "Sacred and Secular Models of Psychology and Religion." *Journal of Psychology and Theology* 5: 197–208.

Carter, John D., and Richard J. Mohline. 1976. "The Nature and Scope of Integration: A Proposal." *Journal of Psychology and Theology* 4: 3–14.

Carter, John D., and Bruce Narramore. 1979. *The Integration of Psychology and Theology: An Introduction*. Grand Rapids: Zondervan.

Catholic Church. 2012. "In the Image of God." *Catechism of the Catholic Church*. (2nd ed.) Vatican City: Libreria Editrice Vaticana. www.vatican.va.

Clarke, Henry. 1835. *Christian Phrenology; or the Teachings of the New Testament Respecting the Animal, Moral, and Intellectual Nature of Man*. Dundee, UK: Advertiser Office.

Clarke, Peter G. H. 2015. *All in the Mind? Does Neuroscience Challenge Faith?* Oxford: Lion Hudson.

Clements, Ronald E. 2000. "The Sources of Wisdom." In *Understanding Wisdom: Sources, Science, and Society*, edited by W. S. Brown, 15–34. West Conshohocken, PA: Templeton Foundation Press.

Clinton, William, and Hillary Clinton. 1997. *The White House Conference on Early Childhood Development and Learning: What New Research on the Brain Tells Us about Our Youngest Children*. April 17. clinton-whitehouse4.archives.gov.

Collicutt, Joanna. 2012. "Bringing the Academic Discipline of Psychology to Bear on the Study of the Bible." *Journal of Theological Studies* 63 (1): 1–48.

Collins, Gary H. 1972. Review of *The Third Force: The Psychology of Abraham Maslow* by Frank Coble. *Science in Christian Perspective*. www.asa3.org.

———. 1977. *The Rebuilding of Psychology: An Integration of Psychology and Christianity*. Wheaton, IL: Tyndale.

———. 1981. *Psychology and Theology: Prospects for Integration*. Nashville: Abingdon Press.

———. 2000. "An Integration View." In *Psychology and Christianity: Four Views*, edited by Eric L. Johnson and Stanton L. Jones, 102–29. Downers Grove, IL: InterVarsity Press.

Combe, George. 1824. *Elements of Phrenology*. London: Simpkin and Marshall.

Cooper, John W. 2000. *Body, Soul, and Life Everlasting: Biblical Anthropology and the Monism-Dualism Debate*. Grand Rapids: Eerdmans.

Copan, Victor. 2016. *Changing Your Mind: The Bible, the Brain, and Spiritual Growth*. Eugene, OR: Cascade Books.

Copernicus, Nicolaus. 1543. *De Revolutionibus Orbium Coelestium*. Nuremberg: Johannes Petreius.

Cosmas. 1897. *The Christian Topography of Cosmas, an Egyptian Monk*.

Edited by J. W. McCrindle. Translated from Greek. London: Hakluyt Society.

Cowan, Charles. 1841. *Phrenology Consistent with Science and Revelation.* London: Sherwood and Co.

Cranefield, Paul F. 1961. "A Seventeenth-Century View of Mental Deficiency and Schizophrenia: Thomas Willis on 'Stupidity or Foolishness.'" *Bulletin of the History of Medicine* 35: 291–316.

Crick, Francis. 1994. *The Astonishing Hypothesis: The Scientific Search for the Soul.* New York: Charles Scribner's Sons.

Crisp, Thomas M., Steven L. Porter, and Gregg A. Ten Elshof. 2016. *Neuroscience and the Soul: The Human Person in Philosophy, Science, and Theology.* Grand Rapids: Eerdmans.

Damasio, Antonio R. 1994. *Descartes' Error: Emotion, Reason, and the Human Brain.* New York: Avon Books.

Darwin, Charles. 1859. *On the Origin of Species: By Means of Natural Selection or the Preservation of Favored Races in the Struggle for Life.* London: John Murray.

———. 1871. *The Descent of Man, and Selection in Relation to Sex.* London: John Murray.

———. 1879. *Darwin Correspondence Database* (entry 12041). www.darwinproject.ac.uk/letter/DCP-LETT-12041.xml.

Davies, Gaius. 2001. *Genius, Grief, and Grace: A Doctor Looks at Suffering and Success.* Fearn, UK: Christian Focus Publications.

———. 2004. "Spiritual Awareness, Personality, and Illness." In *From Cells to Souls—and Beyond*, edited by M. A. Jeeves, 123–45. Grand Rapids: Eerdmans.

Davis, Edward B. 2006. "The Word and the Works: Concordism in American Evangelical Thought." In *The Book of Nature in Early Modern and Modern History*, edited by K. V. Berkel and A. J. Vanderjagt, 196–207. Dudley, MA: Peeters Publishers.

Dax, Marc. 1865. "Observations Tending to Prove the Constant Coincidence of Disturbances of Speech with a Lesion of the Left Hemisphere of the Brain and Lesions of the Left Half of the Encephalon Coincident with the Forgetting of Signs of Thinking." *Bulletin Hebdomadaire de Médecine et de Chirurgie* 2: 259–62. (Published posthumously)

de Felipe, Pablo. 2012. "The Antipodeans and Science and Faith Relations: The Rise, Fall and Vindication of Augustine." In *Augustine beyond the Book: Intermediality, Transmediality, and Reception,* edited by K. Pollmann and M. J. Gill, 281–311. Leiden, Netherlands: Brill.

———. 2017. "Curiosity in the Early Christian Era." Paper presented at the Christians in Science Conference, Oxford. October 21.

de Felipe, Pablo, and Malcolm A. Jeeves. 2017. "Science and Christianity Conflicts: Real and Contrived." *Perspectives on Science and Christian Faith* 69 (3): 131–48.

de Waal, Frans B. M. 1996. *Good Natured: The Origins of Right and Wrong in Humans and Other Animals.* Cambridge, MA: Harvard University Press.

Dienes, Zoltan P., and Malcolm A. Jeeves. 1965. "Thinking in Structures." In *Psychological Monographs on Cognitive Processes.* London: Hutchinson Educational Publishers.

———. 1970. "The Effects of Structural Relations on Transfer." In *Psychological Monographs on Cognitive Processes.* London: Hutchinson Educational Publishers.

di Pellegrino, G., L. Fadiga, L. Fogassi, V. Gallese, and G. Rizzolatti. 1992. "Understanding Motor Events: A Neurophysiological Study." *Experimental Brain Research* 91 (1): 176–80.

Di Vito, Robert A. 1999. "Old Testament Anthropology and the Construction of Personal Identity." *Catholic Biblical Quarterly* 61 (2): 217–38.

Dostoyevsky, Fyodor. 1969. *The Idiot.* Translated by Henry and Olga Carlisle. New York: Signet Classics.

Drake, Stillman. 1980. *Galileo.* Oxford: Oxford University Press.

Draper, John W. 1875. *History of the Conflict between Religion and Science.* New York: D. Appleton and Company.

Easton, William. 1867. *The Harmony of Phrenology and Scripture on the Doctrine of the Soul.* Edinburgh: R. Collie and Son.

Edwards, Jonathan. 1821. *A Treatise Concerning Religious Affections.* Philadelphia: James Crissy.

Eisenberger, Naomi I., Matthew D. Lieberman, and Kipling D. Williams. 2003. "Does Rejection Hurt? An fMRI Study of Social Exclusion." *Science* 302 (5643): 290–92.

Enns, Peter. 2012. *The Evolution of Adam: What the Bible Does and Doesn't Say about Human Origins*. Grand Rapids: Brazos Press.

Erikson, Erik H. 1950. *Childhood and Society*. New York: Norton.

———. 1968. *Identity: Youth and Crisis*. New York: Norton.

Evans, C. Stephen. 1977. *Preserving the Person: A Look at the Human Sciences*. Downers Grove, IL: InterVarsity Press.

———. 2012. "Doing Psychology as a Christian: A Plea for Wholeness." *Journal of Psychology and Theology* 40 (1): 37–40.

Farrer, A. 1964. *Saving Belief*. London: Hodder and Stoughton.

Fernandez-Duque, Diego, Jessica Evans, Colton Christian, and Sara Hodges. 2015. "Superfluous Neuroscience Information Makes Explanations of Psychological Phenomena More Appealing." *Journal of Cognitive Neuroscience* 27 (5): 926–44.

Ferrier, David. 1876. *The Functions of the Brain*. New York: G. P. Putnam's Sons.

Fincham, Frank D. 2013. "I Say a Little Prayer for You: Do Prayers Matter in Marriage and Family Life?" In *Contemporary Issues in Family Studies: Global Perspectives on Partnerships, Parenting, and Support in a Changing World*, edited by A. Abela and J. Walker, 341–45. New York: Wiley.

Fincham, Frank D., Nathaniel M. Lambert, and Steven R. H. Beach. 2010. "Faith and Unfaithfulness: Can Praying for Your Partner Reduce Infidelity?" *Journal of Personality and Social Psychology* 99 (4): 649–59.

Flourens, J. Pierre. 1824. *Recherches Expérimentales sur les Propriétés et les Fonctions du Système Nerveux* (Experimental Research on the Properties and Functions of the Nervous System). Paris: Crevot.

Fowler, Orson. 1844. *Religion: Natural and Revealed; or the Natural Theology and Moral Bearings of Phrenology and Physiology*. New York: Fowler and Wells.

Freud, Sigmund. 1919. *Totem and Taboo: Resemblances between the Psychic Lives of Savages and Neurotics*. London: George Routledge & Sons.

———. 1934. *The Future of an Illusion*. London: Hogarth Press.

———. 1939a. *Civilization and Its Discontents*. London: Hogarth Press.

———. 1939b. *Moses and Monotheism*. London: Hogarth Press.

———. 1949. *An Outline of Psycho-Analysis*. London: Hogarth Press.

Fritsch, Gustav, and Eduard Hitzig. 1870. Über *die Elektrische Erregbarkeit des Grosshirns* [On the electrical excitability of the cerebrum]. *Archiv für Anatomie, Physiologie, und wissenschaftliche Medizin* 37: 300–32.

Fromm, Erich. 1964. *The Heart of Man: Its Genius for Good and Evil*. London: Routledge and Kegan Paul.

Furrow, James L., and Linda M. Wagener. 2000. "Lessons Learned: The Role of Religion in the Development of Wisdom in Adolescence." In *Understanding Wisdom*, edited by W. S. Brown, 361–92. West Conshohocken, PA: Templeton Foundation Press.

Gall, Franz J. 1835a. *Manual of Phrenology: Being an Analytical Summary of the System of Doctor Gall on the Faculties of Man and the Functions of the Brain*. (Translated from the 4th French edition). Philadelphia: Carey, Lea, and Blanchard.

———. 1835b. *On the Functions of the Brain and of Each of Its Parts: With Observations on the Possibility of Determining the Instincts, Propensities and Talents, or the Moral and Intellectual Dispositions of Men and Animals, by the Configuration of the Brain and Head*. Translated by Winslow Lewis. Boston: Marsh, Capen, and Lyon.

Gallese, Vittorio, and Alvin Goldman. 1998. "Mirror Neurons and the Simulation Theory of Mind-Reading." *Trends in Cognitive Sciences* 2 (12): 493–501.

Gallup Organization. 2017. *WIN/Gallup International Poll on Religion in the World*. Washington, DC: Gallup Organization.

Galton, Francis. 1872. "Statistical Inquiries into the Efficacy of Prayer." *Fortnightly Review* 12: 125–35.

Gardner, Howard. 1985. *The Mind's New Science: A History of the Cognitive Revolution*. New York: Basic Books.

Gesch, C. Bernard, Sean Hammond, Sarah Hampson, Anita Eves, and Martin Crowder. 2002. "Influence of Supplementary Vitamins, Minerals and Essential Fatty Acids on the Antisocial Behavior of Young Adult Prisoners." *British Journal of Psychiatry* 181 (1): 22–28.

Gilson, Étienne. 1957. *The Christian Philosophy of St. Thomas Aquinas.* Translated by L. K. Shook. London: Victor Gollanz.

Gould, Stephen Jay. 1997. "Nonoverlapping Magisteria." *Natural History* 106: 16–22.

———. 1999. *Rocks of Ages: Science and Religion in the Fullness of Life.* New York: Ballantine Books.

Grabowski, Thomas, and Antonio R. Damasio. 1996. "Improving Functional Imaging Techniques: The Dream of a Single Image for a Single Mental Event." *Proceedings of the National Academy of Sciences* 93 (25): 14302–3.

Granada, Miguel A. 2008. "Tycho Brahe, Caspar Peucer, and Christoph Rothmann on Cosmology and the Bible." In *Nature and Scripture in the Abrahamic Religions: Up to 1700,* edited by J. van der Meer and S. Mandelbrote, 563–83. Leiden, Netherlands: Brill.

Green, Joel B. 2004. "What Does It Mean to Be Human? Another Chapter in the Ongoing Interaction of Science and Scripture." In *From Cells to Souls and Beyond: Changing Portraits of Human Nature,* edited by M. Jeeves, 179–98. Grand Rapids: Eerdmans.

———. 2008. *Body, Soul, and Human Life: The Nature of Humanity in the Bible.* Grand Rapids: Baker Academic.

———. 2015. *Conversion in Luke–Acts: Divine Action, Human Cognition, and the People of God.* Grand Rapids: Baker Academic.

Greene, Joshua D., Leigh E. Nystrom, Andrew D. Engell, John M. Darley, and Jonathan D. Cohen. 2004. "The Neural Bases of Cognitive Conflict and Control in Moral Judgment." *Neuron* 44 (2): 389–400.

Greene, Joshua D., Brian Sommerville, Leigh E. Nystrom, John M. Darley, and Jonathan D. Cohen. 2001. "An fMRI Investigation of Emotional Engagement in Moral Judgment." *Science* 293 (5537): 2105–8.

Groopman, Jeremy. 2001. "God on the Brain." *The New Yorker,* September 17, 165–68.

Hakeem, A. Y., C. C. Sherwood, C. J. Bonar, C. Butti, P. R. Hof, and J. M. Allman. 2009. "Von Economo Neurons in the Elephant Brain." *Anatomical Record* 292 (2): 242–48.

Hall, M. Elizabeth Lewis, Richard L. Gorsuch, H. Newton Malony, S. Bruce

Narramore, and Mary S. Van Leeuwen. 2006. "Dialogue, Embodiment, and the Unity of Faith and Learning: A Conversation on Integration in a Postmodern Age." *Journal of Psychology and Christianity* 25 (4): 331–37.

Harrison, Peter. 2006."The 'Book of Nature' and Early Modern Science." In *The Book of Nature in Early Modern and Modern History*, edited by K. van Berkel and A. Vanderjagt, 1–26. Leuven, Belgium: Peeters Publishers.

———, ed. 2010. *The Cambridge Companion to Science and Religion*. New York: Cambridge University Press.

———. 2015. *The Territories of Science and Religion*. Chicago: University of Chicago Press.

Hasker, William. 2010. "On Behalf of Emergent Dualism." In *In Search of the Soul: Perspectives on the Mind-Body Problem*, edited by J. B. Green, 75–100. Eugene, OR: Wipf and Stock Publishers.

Hawkes, Nick. 2012. *Evidence of God: A Scientific Case for God*. Eugene, OR: Wipf & Stock Publishers.

Hearnshaw, Leslie Spencer. 1964. *A Short History of British Psychology: 1840–1940*. New York: Barnes & Noble.

Holeman, Virginia. 2004. "The Neuroscience of Christian Counseling?" In *What About the Soul? Neuroscience and Christian Anthropology*, edited by J. B. Green, 145–58. Nashville: Abingdon Press.

Hubel, D. H. 1979. "The Brain." *Scientific American* 241 (3): 45–53.

Hubel, David H., and Torsten N. Wiesel. 1962. "Receptive Fields, Binocular Interaction and Functional Architecture in the Cat's Visual Cortex." *Journal of Physiology* 160 (1): 106–54.

Huxley, Thomas H. 1860. "Darwin on the Origin of Species." *Westminster Review* 17: 541–70.

Jackson, Hughlings. 1899. "Epileptic Attacks with a Warning of a Crude Sensation of Smell and with the Intellectual Aura (Dreamy State) in a Patient Who Had Symptoms Pointing to Gross Organic Disease of the Right Temporo-Sphenoidal Lobe." *Brain* 22 (4): 535–49.

James, William. 1890. *The Principles of Psychology*, Vol. 2. New York: Holt.

———. 1902. *The Varieties of Religious Experience: A Study of Human*

*Nature*. New York: Longmans, Green, and Company. (Gifford Lectures on natural religion delivered at Edinburgh in 1901–2)

Jeeves, Malcolm A. 1953. "Wishful Un-Thinking." In *Where Science and Faith Meet*, contributors James B. Torrance, Donald M. MacKay, Malcolm Jeeves, Robert L. F. Boyd, and Oliver Barclay, 21–26. London: InterVarsity Press.

———. 1965. "Psychological Studies of Three Cases of Agenesis of the Corpus Callosum." In *Functions of the Corpus Callosum*, edited by E. G. Ettlinger, 77–94. London: J. and A. Churchill.

———. 1976. *Psychology and Christianity: The View Both Ways*. Downers Grove, IL: InterVarsity Press.

———. 1988. "The Psychology of Religion." In *New Dictionary of Theology*, edited by S. B. Ferguson, D. F. Wright, and J. I. Packer, 543–47. Downers Grove, IL: InterVarsity Press.

———. 1994. *Mind Fields*. Grand Rapids: Baker Book House.

———. (1997) 2006. *Human Nature: Reflections on the Integration of Psychology and Christianity*. Grand Rapids: Baker Books. Reprint, West Conshohocken, PA: Templeton Press.

———, ed. 2004. *From Cells to Souls—and Beyond: Changing Portraits of Human Nature*. Grand Rapids: Eerdmans.

———, ed. 2011. *Rethinking Human Nature: A Multidisciplinary Approach*. Grand Rapids: Eerdmans.

———. 2013. "Psychologizing and Neurologizing about Religion: Facts, Fallacies, and the Future." In *Science and Religion in the Twenty-First Century*, edited by R. R. Manning and M. Byrne, 75–93. London: SCM Press.

———. 2015a. "Afterword: Toward an Emerging Picture of How We Came to Be What We Seem to Be." In *The Emergence of Personhood: A Quantum Leap?* edited by M. A. Jeeves, 220–41. Grand Rapids: William B. Eerdmans.

———, ed. 2015b. *The Emergence of Personhood: A Quantum Leap?* Grand Rapids: William B. Eerdmans.

———. 2017. "The Testimony of the Spirit: Insights from Psychology and Neuroscience." In *The Testimony of the Spirit: New Essays*, edited by R.

Douglas Geivett and Paul K. Moser, 127–50. New York: Oxford University Press.

Jeeves, Malcolm A., and Günter Baumgartner. 1986. *Methods in Neuropsychology.* Oxford: Pergamon Press.

Jeeves, Malcolm A., and R. J. Berry. 1998. *Science, Life, and Christian Belief: A Survey and Assessment.* Leicester, UK: InterVarsity Press.

Jeeves, Malcolm A., and Warren S. Brown. 2009. *Neuroscience, Psychology, and Religion: Illusions, Delusions, and Realities about Human Nature.* West Conshohocken, PA: Templeton Press.

Jeeves, Malcolm A., and Brian G. Greer. 1983. *Analysis of Structural Learning.* London: Academic Press.

Jeeves, Malcolm A., T. E. Ludwig, P. Moes, and W. D. Norman. 2001. "The Stability of Compromised Interhemispheric Processing in Callosal Dysgenesis and Partial Commissurotomy." *Cortex* 37 (5): 643–64.

Jeeves, Malcolm A., and A. H. Winefield. 1969. "Discrimination-Reversal Skills in Squirrel Monkeys." *Psychonomic Science* 14 (5): 222–23.

Johnson, Dale A. 2017. *Newton in the Pulpit.* Chongqing, China: New Sinai Press.

Johnson, Eric L., ed. 2010. *Psychology & Christianity: Five Views.* (2nd ed.) Downers Grove, IL: InterVarsity Press.

———. 2011. "The Three Faces of Integration." *Journal of Psychology and Christianity* 30 (4): 339–55.

Johnson, Eric L., and Stanton L. Jones, ed. 2000. *Psychology and Christianity: Four Views.* Downers Grove, IL: InterVarsity Press.

Johnson, Mark. 2000. "State of the Art: How Babies' Brains Work." *The Psychologist* 13 (June): 298–301.

Johnson, Mark, Michelle de Haan, Andrew Oliver, Warwick Smith, Haralambos Hatzakis, Leslie Tucker, and G. Csibra. 2001. "Recording and Analyzing High-Density Event-Related Potentials with Infants Using the Geodesic Sensor Net." *Developmental Neuropsychology* 19 (3): 295–323.

Johnson, Micah. 2018. "How Responsible Are Killers with Brain Damage?" *Scientific American,* January 30, 2018. www.scientificamerican.com /article/how-responsible-are-killers-with-brain-damage/

Johnson, Samuel. 1825. *The Works of Samuel Johnson: Sermons*. Oxford: Talboys and Wheeler.

Jones, Jeffrey M. 2010. "Few Americans Oppose National Day of Prayer." In *The Gallup Poll: Public Opinion 2010*, edited by Frank Newport, 150–51. Lanham, MD: Rowman and Littlefield.

Jones, Stanton, ed. 1986. *Psychology and the Christian Faith: An Introductory Reader*. Grand Rapids: Baker Publishing Group.

———. 1994. "A Constructive Relationship for Religion with the Science and Profession of Psychology: Perhaps the Boldest Model Yet." *American Psychologist* 49 (3): 184–99.

———. 2010. "An Integration View." In *Psychology & Christianity: Five Views*. (2nd ed.) Edited by Eric L. Johnson, 101–28. Downers Grove, IL: InterVarsity Press.

Jung, Carl G. 1968. *The Archetypes and the Collective Unconscious*. (2nd ed.) Abingdon, UK: Routledge.

Kahn, Chris. 2003. "Tumor Goes, So Does Bad Behavior." *CBS News*, July 23. www.cbsnews.com.

Kanwisher, Nancy. 2000. "Domain Specificity in Face Perception." *Nature Neuroscience* 3 (8): 759–63.

Kanwisher, Nancy, and Galit Yovel. 2006. "The Fusiform Face Area: A Cortical Region Specialized for the Perception of Faces." *Philosophical Transactions of the Royal Society* 361 (1476): 2109–28.

Kark, Jeremy, Galia Shemi, Yechiel Friedlander, Oz Martin, Orly Manor, and S. H. Blondheim. 1996. "Does Religious Observance Promote Health? Mortality in Secular vs. Religious Kibbutzim in Israel." *American Journal of Public Health* 86 (3): 341–46.

Karmiloff-Smith, Annette. 2009. "Nativism versus Neuroconstructivism: Rethinking the Study of Developmental Disorders." *Developmental Psychology* 45 (1): 56–63.

Kendell, Robert E. 2001. "The Distinction between Mental and Physical Illness." *British Journal of Psychiatry* 178: 490–93.

Keysers, Christian, and David Perrett. 2004. "Demystifying Social Cognition: A Hebbian Perspective." *Trends in Cognitive Sciences* 8 (11): 501–7.

Kierkegaard, Søren. 1944. *For Self-Examination / Judge for Yourselves.* Translated by W. Lowrie. Princeton, NJ: Princeton University Press.

Kilner, J. M., and R. N. Lemon. 2013. "What We Know Currently about Mirror Neurons." *Current Biology* 23 (23): R1057–62.

Kingsley, Charles. 1859. *Darwin Correspondence Database* (entry 2534). www.darwinproject.ac.uk/letter/DCP-LETT-2534.xml.

Koteskey, Ronald L. 1980. *Psychology from a Christian Perspective.* Nashville: Abingdon Press.

Künkel, Fritz. 1954. "The Integration of Religion and Psychology." *Journal of Psychotherapy as a Religious Process* 1: 1–11.

Lambert, Heath. 2016. *A Theology of Biblical Counseling: The Doctrinal Foundations of Counseling Ministry.* Grand Rapids: Zondervan.

Larner, Andrew, and John P. Leach. 2002. "Phineas Gage and the Beginnings of Neuropsychology." *Advances in Clinical Neuroscience and Rehabilitation* 2 (3): 26.

Lashley, Karl. 1929. *Brain Mechanisms and Intelligence.* Chicago: University of Chicago Press.

Lassonde, Maryse, and Malcolm A. Jeeves, ed. 1994. *Callosal Agenesis: A Natural Split Brain?* New York: Plenum Press.

Lemke, Steve, John Walton, and Kenneth Samples. 2017. "Biblical Interpretation: What Is the Nature of Biblical Authority?" In *Old-Earth or Evolutionary Creation?* edited by K. Keathley, J. B. Stump, and J. Aguirre, 27–48. Downers Grove, IL: InterVarsity Press.

Leuba, James Henry. 1912. *A Psychological Study of Religion: Its Origin, Function, and Future.* New York: Macmillan.

Lewis, C. S. 1947. *Miracles: How God Intervenes in Nature and Human Affairs.* New York: Macmillan.

———. 1981. *Christian Reflections.* Glasgow, UK: Collins.

Lewis, Tanya. 2016. "How Dopamine Tunes Working Memory." *The Scientist*, June 3. www.the-scientist.com.

Lindberg, David C., and Ronald L. Numbers. 1987. "Beyond War and Peace: A Reappraisal of the Encounter between Christianity and Science." *Perspectives on Science and Christian Faith* 39 (3): 140–49.

Loftus, Elizabeth F., and John C. Palmer. 1974. "Reconstruction of Automobile Destruction: An Example of the Interaction between Language and Memory." *Journal of Verbal Learning and Verbal Behavior* 13 (5): 585–89.

Lossky, Vladimir. 1991. *The Mystical Theology of the Eastern Church.* Cambridge: Clarke.

Ludwig, Thomas E. 1982. "Age Differences in Mental Synthesis." *Journal of Gerontology* 37 (2): 182–89.

———. (1986, 1989, 1992, 1996, 2004) 2015. *PsychSim: Interactive Graphic Simulations for Psychology.* New York: Worth Publishers.

———. 1997. "Selves and Brains: Tracing a Path between Interactionism and Materialism." *Philosophical Psychology* 10 (4): 489–95.

———. 2002. *PsychInquiry: Student Activities in Research and Critical Thinking.* New York: Worth Publishers.

Ludwig, Thomas E., and M. A. Jeeves. 1996. "Maximizing the Bilateral Field Advantage on Verbal and Nonverbal Matching Tasks." *Cortex* 32 (1): 131–41.

Ludwig, Thomas E., and C. W. Perdue. 2005. "Multimedia and Computer-Based Learning in Introductory Psychology." In *Best Practices for Teaching Introduction to Psychology*, edited by D. S. Dunn and S. L. Chew, 143–58. Mahwah, NJ: Erlbaum.

Luther, Martin. 1955. *Letters of Spiritual Counsel.* Translated by Theodore G. Tappert. Philadelphia: Westminster Press.

MacKay, Donald M. 1953. "Looking for Connections." In *Where Science and Faith Meet*, contributors James B. Torrance, Donald M. MacKay, Malcolm Jeeves, Robert L. F. Boyd, and Oliver Barclay, 14–20. London: InterVarsity Press.

———. 1960. *Science and Christian Faith Today.* London: Falcon.

———. 1974. *The Clockwork Image: A Christian Perspective on Science.* London: InterVarsity Press.

———. 1978. "Selves and Brains." *Neuroscience* 3 (7): 599–606.

———. 1991. *Behind the Eye.* Edited by Valerie MacKay. Oxford: Basil Blackwell.

MacKay, Donald M., and T. E. Ludwig. 1986. "Source-Density Analysis of

Scalp Potentials during Linguistic and Non-Linguistic Judgments." *Experimental Brain Research* 64: 434–46.

Macmillan, Malcolm, and Matthew L. Lena. 2010. "Rehabilitating Phineas Gage." *Neuropsychological Rehabilitation* 20 (5): 641–58.

Maguire, Eleanor, David Gadian, Ingrid Johnsrude, Catriona Good, John Ashburner, Richard Frackowiak, and Christopher Frith. 2000. "Navigation-Related Structural Change in the Hippocampi of Taxi Drivers." *Proceedings of the National Academy of Sciences* 97 (8): 4398–403.

Malinowski, Bronislaw. 1927. *Sex and Repression in Savage Society.* London: Kegan Paul, Trench, Trubner & Co.

———. 1936. *The Foundations of Faith and Morals: An Anthropological Analysis of Primitive Beliefs and Conduct with Special Reference to the Fundamental Problems of Religion and Ethics.* London: Oxford University Press.

Marsden, George M. 2003. *Jonathan Edwards: A Life.* New Haven, CT: Yale University Press.

Maslow, Abraham H. 1968. *Toward a Psychology of Being.* (2nd ed.) New York: Van Nostrand Reinhold.

McCall Smith, Alexander. 2004. "Human Action, Neuroscience and Law." In *The New Brain Sciences: Perils and Prospects*, edited by D. Rees and S. Rose, 103–22. Cambridge: Cambridge University Press.

McClendon, B. T., and S. Prentice-Dunn. 2001. "Reducing Skin Cancer Risk: An Intervention Based on Protection Motivation Theory." *Journal of Health Psychology* 6 (3): 321–28.

McGrath, Alister. 2015. "Conflict or Mutual Enrichment? Why Science and Theology Need to Talk to Each Other." *Science and Christian Belief* 27 (1): 3–16.

McMullin, Ernan, ed. 2005. *The Church and Galileo.* Notre Dame, IN: University of Notre Dame Press.

———. 2009. "The Galileo Affair." *Faraday Papers* no. 17: 227–32. Cambridge: Faraday Institute.

McNamara, Patrick, Raymon Durso, and Ariel Brown. 2006. "Religiosity in Patients with Parkinson's Disease." *Neuropsychiatric Disease and Treatment* 2 (3): 341–48.

Meehl, Paul E., Richard Klann, Alfred Schmieding, Kenneth Breimeier, and Sophie Schroeder-Slomann. 1958. *What, Then, Is Man? A Symposium of Theology, Psychology, and Psychiatry.* St. Louis, MO: Concordia Publishing House.

Middleton, J. Richard. 2005. *The Liberating Image: The Imago Dei in Genesis 1.* Grand Rapids: Brazos Press.

Milner, A. David, and Michael Rugg. 1995. "Neuropsychological and Developmental Studies of the Corpus Callosum." *Neuropsychologia* 33 (8): 921–1007. (Special issue to mark the retirement of Malcolm A. Jeeves, CBE, formerly editor in chief of *Neuropsychologia*)

Miner, Maureen, and Martin Dowson. 2012. "Spiritual Experiences Reconsidered: A Relational Approach to the Integration of Psychology and Theology." *Journal of Psychology and Theology* 40 (1): 55–59.

Moes, Paul, and Donald J. Tellinghuisen. 2014. *Exploring Psychology and Christian Faith: An Introductory Guide.* Grand Rapids: Baker Academic.

Munn, Norman L. 1946. *Psychology: The Fundamentals of Human Adjustment.* Boston: Houghton Mifflin.

———. 1980. *Being and Becoming: An Autobiography.* Adelaide, South Australia: Adelaide University Union Press.

Myers, David G. 1978. *The Human Puzzle: Psychological Research and Christian Belief.* San Francisco: Harper & Row.

———. 1986. "Social Psychology." In *Psychology and the Christian Faith: An Introductory Reader,* edited by Stanton Jones, 217–39. Grand Rapids: Baker.

———. 1987. "Yin and Yang in Psychological Research and Christian Belief." *Perspectives in Science and Christian Faith* 39 (3): 128–39.

———. 1998. "Psychology, Applied Spirituality and Health: Do They Relate?" *Science and Spirit* 9: 30.

———. 2008. *A Friendly Letter to Skeptics and Atheists: Musings on Why God Is Good and Faith Isn't Evil.* San Francisco: Jossey-Bass.

———. 2010a. "A Levels of Explanation View." In *Psychology and Christianity: Five Views,* edited by E. L. Johnson, 49–78. (2nd ed.) Downers Grove, IL: InterVarsity Press.

———. 2010b. "A Levels-of-Explanation Response to Christian Psychology." In *Psychology and Christianity: Five Views,* edited by E. L. Johnson, 179–82. (2nd ed.) Downers Grove, IL: InterVarsity Press.

———. 2010c. "A Levels-of-Explanation Response to Integration." In *Psychology and Christianity: Five Views,* edited by E. L. Johnson, 129–31. (2nd edition) Downers Grove, IL: InterVarsity Press.

———. 2010d. "A Levels-of-Explanation Response to Biblical Counseling." In *Psychology and Christianity: Five Views* (2nd edition), edited by E. L. Johnson, 274–5. (2nd ed.) Downers Grove, IL: InterVarsity Press.

———. 2015. "Psychological Science Meets Christian Faith." In *Reason and Wonder: Why Science and Faith Need Each Other,* edited by E. Priest, 126–40. London: SPCK Publishing.

Myers, David G., and C. Nathan DeWall. 2018. *Psychology.* (12th ed.) New York: Worth Publishers.

Myers, David G., and Malcolm A. Jeeves. (1987) 2003. *Psychology through the Eyes of Faith.* San Francisco: Harper.

Naito, H., and N. Matsui. 1988. "Temporal Lobe Epilepsy with Ictal Ecstatic State and Interictal Behavior of Hypergraphia." *Journal of Nervous and Mental Disease* 176 (2): 123–24.

Nichols, David E., and Benjamin R. Chemel. 2006. "The Neuropharmacology of Religious Experience: Hallucinogens and the Experience of the Divine." In *Where God and Science Meet*: Vol. 3, *The Psychology of Religious Experience,* edited by P. McNamara, 1–33. Westport, CT: Praeger.

Norlie, Olaf M. 1924. *An Elementary Christian Psychology.* Minneapolis: Augsburg Press.

Norman, Wayne D., and Malcolm A. Jeeves. 2010. "Neurotheology: Avoiding a Reinvented Phrenology." *Perspectives on Science and Christian Faith* 62 (4): 235–51.

Numbers, Ronald L, ed. 2009. *Galileo Goes to Jail—and Other Myths about Science and Religion.* Cambridge, MA: Harvard University Press.

O'Craven, Kathleen, and Nancy Kanwisher. 2000. "Mental Imagery of Faces and Places Activates Corresponding Stimulus-Specific Brain Regions." *Journal of Cognitive Neuroscience* 12 (6): 1013–23.

Pannenberg, Wolfhart. 1944. *Systematic Theology*, Vol. 2. Grand Rapids: Eerdmans.

Pascal, Blaise. (1670) 1958. *Pascal's Pensées (Les Pensées)*. New York: Dutton.

Pavlov, Ivan. 1951. *Complete Collected Works*. Vol. 3. Moscow: Academy of Science.

Payer, Doris, Christy Marshuetz, Brad Sutton, Andy Hebrank, Robert C. Welsh, and Denise C. Park. 2006. "Decreased Neural Specialization in Old Adults on a Working Memory Task." *Neuroreport* 17 (5): 487–91.

Penchansky, David. 2012. *Understanding Wisdom Literature*. Grand Rapids: Eerdmans.

Perrett, David I., Edmund T. Rolls, and Woody Caan. 1982. "Visual Neurons Responsive to Faces in the Monkey Temporal Cortex." *Experimental Brain Research* 47 (3): 329–42.

Perrett, David I., P. A. Smith, D. D. Potter, A. J. Mistlin, A. S. Head, A. D. Milner, and M. A. Jeeves. 1984. "Neurons Responsive to Faces in the Temporal Cortex: Studies of Functional Organization, Sensitivity to Identity, and Relation to Perception." *Human Neurobiology* 3 (4): 197–208.

Petrovic, Predrag, Eija Kalso, Karl M. Petersson, and Martin Ingvar. 2002. "Placebo and Opioid Analgesia—Imaging a Shared Neuronal Network." *Science* 295: 1737–40.

Powell, Baden. 1860. "On the Study of the Evidences of Christianity." In *Essays and Reviews*, edited by John William Parker, 94–144. (2nd ed.) London: John W. Parker and Son.

Powlison, D. 2010. "A Biblical Counseling View." In *Psychology and Christianity: Five Views*, edited by E. L. Johnson, 245–62. (2nd ed.) Downers Grove, IL: InterVarsity Press.

Pratt, James Bissett. 1920. *The Religious Consciousness: A Psychological Study*. New York: Macmillan.

Premack, David, and Gary Woodruff. 1978. "Does the Chimpanzee Have a Theory of Mind?" *Behavioral and Brain Sciences* 1 (4): 515–26.

Priest, Eric. ed. 2015. *Reason and Wonder: Why Science and Faith Need Each Other*. London: SPCK Publishing.

Rahner, Karl. 1990. *Faith in a Wintry Season.* Translated by P. Imhof, H. Biallowons, and H. D. Egan. New York: Crossroad.

Rajalakshmi, R., and Malcolm A. Jeeves. 1965. "The Relative Difficulty of Reversal Learning (Reversal Index) as a Basis of Behavioral Comparisons." *Animal Behavior* 13 (2): 203–11.

Ramachandran, Vilayanur S. 2000. "Mirror Neurons and Imitation Learning as the Driving Force behind the Great Leap Forward in Human Evolution." *The Edge,* May 31. edge.org.

Ramachandran, Vilayanur S., and Sandra Blakeslee. 1998. *Phantoms in the Brain: Probing the Mysteries of the Human Mind.* New York: William Morrow.

Ramachandran, Vilayanur S., W. Hirstein, K. C. Armel, E. Tecoma, and V. Iragul. 1997. "The Neural Basis of Religious Experience." Paper presented at the 27th annual meeting of the Society for Neuroscience, New Orleans, LA. October 25–30.

Ramakrishnan, Venkatraman. 2016. "Scientific Insight." *The Times* (London). March 5

Rambo, Lewis R., and Charles E. Farhadian, eds. 2014. *The Oxford Handbook of Religious Conversion.* New York: Oxford University Press.

Randles, W. G. L. 1994. "Classical Models of World Geography and Their Transformation Following the Discovery of America." In *The Classical Tradition and the Americas.* Vol. 1: *European Images of the Americas and the Classical Tradition,* edited by W. Haase and M. Reinhold, 5–76. New York: Walter de Gruyter.

Redfern, Clare, and Alasdair Coles. 2015. "Parkinson's Disease, Religion, and Spirituality." *Movement Disorders Clinical Practice* 2 (4): 341–46.

Redman, Judith C. S. 2010. "How Accurate Are Eyewitnesses? Bauckham and the Eyewitnesses in the Light of Psychological Research." *Journal of Biblical Literature* 129 (1): 177–97.

Rees, Dai, and Barbro Westerholm. 2004. "Conclusion." In *The New Brain Sciences: Perils and Prospects,* edited by D. Rees and S. Rose, 265–75. Cambridge: Cambridge University Press.

Roberts, Robert. 2012. "The Idea of a Christian Psychology." *Journal of Psychology and Theology* 40 (1): 37–40.

Roberts, Robert, and P. J. Watson. 2010. "A Christian Psychology View." In *Psychology and Christianity: Five Views,* edited by E. L. Johnson, 149–78. (2nd ed.) Downers Grove, IL: InterVarsity Press.

Rogers, Carl R. 1951. *Client-Centered Therapy.* Boston: Houghton Mifflin.

———. 1957. "The Necessary and Sufficient Conditions of Therapeutic Personality Change." *Journal of Consulting Psychology* 21: 95–103.

———. 1980. *A Way of Being.* Boston: Houghton Mifflin.

Rumbaugh, D. M., and Malcolm A. Jeeves. 1966. "A Comparison of Two Discrimination-Reversal Indices Intended for Use with Diverse Groups of Organisms." *Psychonomic Science* 6: 1–2.

Russell, Colin A. 1989. "The Conflict Metaphor and Its Social Origins." *Science and Christian Belief* 1 (1): 3–26.

Sadato, Norihiro, Alvaro Pascual-Leone, Jordan Grafman, Deiber Ibañez, Marie-Pierre Vincente, George Dold, and Mark Hallett. 1996. "Activation of the Primary Visual Cortex by Braille Reading in Blind Subjects." *Nature* 380 (6574): 526–28.

Salzman, Mark. 2001. *Lying Awake.* New York: Knopf Doubleday.

Sargant, William. 1957. *Battle for the Mind: A Physiology of Conversion and Brain-Washing.* New York: Doubleday and Company.

———. 1974. *The Mind Possessed: A Physiology of Possession, Mysticism, and Faith Healing.* Philadelphia: Lippincott.

Saver, Jeffrey L., and John Rabin. 1997. "The Neural Substrates of Religious Experience." *Journal of Neuropsychiatry* 9 (3): 498–510.

Schloss, Jeffrey P. 2000. "Wisdom Traditions as Mechanisms for Organizational Integration: Evolutionary Perspectives on Homeostatic 'Laws of Life.'" In *Understanding Wisdom: Sources, Science, and Society,* edited by W. S. Brown, 154–91. West Conshohocken, PA: Templeton Press.

Scott, Rodney J., and Raymond E. Phinney, Jr. 2012. "Relating Body and Soul: Insights from Development and Neurobiology." *Perspectives on Science and Christian Faith* 64 (2): 90–107.

Scott, William. 1837. *The Harmony of Phrenology with Scripture.* Edinburgh: Fraser and Co.

Siegel, Zachary. 2015. "America's First Legal Ayahuasca 'Church.'" *Daily Beast,* December 15. www.thedailybeast.com.

Sinkman, Arthur. 2008. "The Syndrome of Capgras." *Psychiatry: Interpersonal and Biological Processes* 71 (4): 371–78.

Sizemore, Timothy. 2016. *The Psychology of Religion and Spirituality.* Hoboken, NJ: Wiley.

Skinner, B. F. 1971. *Beyond Freedom and Dignity.* Indianapolis, IN: Hackett Publishing Company.

Smedes, Lewis B. 2003. *My God and I.* Grand Rapids: Eerdmans.

Sperry, Roger. 1987. "Consciousness and Causality." In *The Oxford Companion to the Mind,* edited by R. L. Gregory, 164–66. Oxford: Oxford University Press.

Spinks, G. Stephens. 1963. *Psychology and Religion: An Introduction to Contemporary Views.* London: Methuen & Co.

Spurzheim, Johann C. 1827. *Outlines of Phrenology.* London: Treuttel, Wurtz, and Richter.

———. 1840. *A View of the Philosophical Principles of Phrenology.* London: Charles Knight.

Starbuck, Edwin Diller. 1899. *The Psychology of Religion: An Empirical Study of the Growth of Religious Consciousness.* London: Walter Scott.

Stawski, Christopher. 2004. "Spiritual Transformation Q&A: Mario Beauregard." *Metanexus,* March 1. metanexus.net.

Stone, Lawson. 2004. "The Soul: Possession, Part, or Person?" In *What about the Soul? Neuroscience and Christian Anthropology,* edited by J. B. Green, 47–61. Nashville: Abingdon Press.

Stott, John R. W. 1994. *The Message of Romans: God's Good News for the World.* Leicester, UK: InterVarsity Press.

Strawn, Brad D., Ronald W. Wright, and Paul Jones. 2014. "Tradition-Based Integration: Illuminating the Stories and Practices That Shape Our Integrative Imagination." *Journal of Psychology and Christianity* 33 (4): 300–10.

Sullivan, J. E. 1963. *The Image of God.* Dubuque, IA: Priory Press.

Thiselton, Anthony C. 2009. *The Living Paul: An Introduction to the Apostle's Life and Thought.* Downers Grove, IL: InterVarsity Press Academic.

———. 2015. *The Thiselton Companion to Christian Theology.* Grand Rapids: Eerdmans.

Thomas (of Aquino). 1912. *The Summa Theologica of St. Thomas Aquinas.* London: Burns Oates and Washbourne.

Thomas, Gary. 1997. "Doctors Who Pray." *Christianity Today* 41 (1): 20–30.

Thomson, Keith S. 1996. "The Revival of Experiments on Prayer." *American Scientist* 84 (6): 532–34.

Thorpe, William H. 1956. *Learning and Instinct in Animals.* London: Methuen.

———. 1974. *Animal Nature and Human Nature.* Garden City, NY: Anchor Press / Doubleday.

Thouless, Robert H. 1923. *An Introduction to the Psychology of Religion.* London: Cambridge University Press.

Tjeltveit, Alan C. 2012. "Lost Opportunities, Partial Success, and Key Questions: Some Historical Lessons." *Journal of Psychology and Theology* 40 (1): 16–20.

Tomasello, Michael. 2000. "Primate Cognition: Introduction to the Issue." *Cognitive Science* 24 (3): 357.

Tomasello, Michael, Josep Call, and Brian Hare. 2003. "Chimpanzees Understand Psychological States—The Question Is Which Ones and to What Extent." *Trends in Cognitive Sciences* 7 (4): 153–56.

Tooby, John, and Leda Cosmides. 1992. "The Psychological Foundations of Culture." In *The Adapted Mind: Evolutionary Psychology and the Generation of Culture,* edited by J. Barkow, L. Cosmides, and J. Tooby, 19–136. New York: Oxford University Press.

Vande Kemp, Hendrika. 1996. "Historical Perspective: Religion and Clinical Psychology in America." In *Religion and the Clinical Practice of Psychology,* edited by Edward Schafranske, 71–112. Washington, DC: American Psychological Association.

———. 1998. "Christian Psychologies for the Twenty-First Century: Lessons from History." *Journal of Psychology and Christianity* 17 (3): 197–209.

Van Horn, John D., Andrei Irimia, Carinna M. Torgerson, Micah C. Chambers, Ron Kikinis, and Arthur W. Toga. 2012. "Mapping Connectivity Damage in the Case of Phineas Gage." *PLos ONE* 7 (5): e37454.

Venema, Dennis, and Scot McKnight. 2017. *Adam and the Genome: Reading Scripture after Genetic Science.* Grand Rapids: Brazos Press.

Vitz, Paul. 1977. *Psychology as Religion: The Cult of Self-Worship*. Grand Rapids: Eerdmans.

Volf, Miroslav. 2005. *Free of Charge: Giving and Forgiving in a Culture Stripped of Grace*. Grand Rapids: Zondervan.

von Rad, Gerhard. 1961. *Genesis: A Commentary*. Translated by John H. Marks. Philadelphia: Westminster Press.

Walton, John H. 2009. *The Lost World of Genesis One: Ancient Cosmology and the Origins Debate*. Downers Grove, IL: InterVarsity Press.

———. 2015. *The Lost World of Adam and Eve: Genesis 2–3 and the Human Origins Debate*. Downers Grove, IL: InterVarsity Press.

Watts, Fraser. 2002. *Theology and Psychology*. Basingstoke, UK: Ashgate.

———. 2010. "Psychology and Theology." In *The Cambridge Companion to Science and Religion*, edited by P. Harrison, 190–206. New York: Cambridge University Press.

———. 2012. "Doing Theology in Dialogue with Psychology." *Journal of Psychology and Theology* 40 (1): 45–50.

Weaver, Glenn. 2004. "Embodied Spirituality: Experiences of Identity and Spiritual Suffering among Persons with Alzheimer's Dementia." In *From Cells to Souls—and Beyond. Changing Portraits of Human Nature*, edited by M. Jeeves, 77–101. Grand Rapids: Eerdmans.

Wellcome Collection. 2016. *Mind & Body*, February. wellcomecollection .org.

Wells, David F. 1989. *Turning to God: Biblical Conversion in the Modern World*. Grand Rapids: Baker.

Wernicke, Carl. 1874. *Der Aphasische Symptomencomplex: Eine Psycholo-gische Studie auf Anatomischer Basis* [The aphasic symptom complex: A psychological study from an anatomical basis]. Breslau, Germany: Max Cohn & Weigert.

Westermann, Claus. 1984. *Genesis 1–11*. Translated by J. J. Scullion. London: SPCK Publishing.

White, Andrew D. 1896. *A History of the Warfare of Science with Theology in Christendom*, Vol. 1. New York: D. Appleton and Company.

Whiten, Andrew. 2006. "The Place of 'Deep Social Mind' in the Evolution of Human Nature." In *Human Nature*, edited by M. A. Jeeves, 207–22. Edinburgh: Royal Society of Edinburgh.

Whiten, Andrew, and Richard Byrne. 1997. *Machiavellian Intelligence II: Extensions and Evaluations.* Cambridge: Cambridge University Press.

Willis, Thomas. 1664. *Cerebri Anatome* [The anatomy of the brain]. London: Martyn and Allestry.

Witvliet, C. V. O., T. E. Ludwig, and K. L. Vander Laan. 2001. "Granting Forgiveness or Harboring Grudges: Implications for Emotion, Physiology, and Health." *Psychological Science* 12 (2): 117–23.

Wixted, John T., and Gary L. Wells. 2017. "The Relationship between Eyewitness Confidence and Identification Accuracy: A New Synthesis." *Psychological Science in the Public Interest* 18 (1): 10–65.

Wolska-Conus, W. 1989. "Stéphanos d'Athènes et Stéphanos d'Alexandrie. Essai d'identification et de biographie." *Revue des études Byzantines* 47: 5–89.

Wolterstorff, Nicholas. 1984. *Reason within the Bounds of Religion.* (2nd ed.) Grand Rapids: Eerdmans.

Woodward, Kenneth L. 1997. "Is God Listening?" *Newsweek*, 56–64. March 31.

Woollett, K., H. J. Spiers, and E. A. Maguire. 2009. "Talent in the Taxi: A Model System for Exploring Expertise." *Philosophical Transactions of the Royal Society of London* 364 (1522): 1407–16.

Worthington, Everett L. 2010. *Coming to Peace with Psychology: What Christians Can Learn from Psychological Science.* Downers Grove, IL: InterVarsity Press Academic.

Worthington, Everett L., and Todd W. Hall. 2012. "Editors' Introduction to the Special Issue." *Journal of Psychology and Theology* 40 (1): 3–4.

Wright, Chris. 2004. *Old Testament Ethics for the People of God.* Leicester, UK: InterVarsity Press.

Wright, N. T. 2004. Endorsement of *From Cells to Souls—and Beyond* (edited by Malcolm Jeeves [Grand Rapids: Eerdmans]), eerdmans.com/Products/0985/from-cells-to-souls---and-beyond.aspx.

———. 2011. *Scripture and the Authority of God: How to Read the Bible Today.* New York: Harper.

Yangarber-Hicks, Natalia, Charles Behensky, Sally S. Canning, Kelly S. Flanagan, Nicholas J. S. Gibson, Mitchell W. Hicks, Cynthia N. Kimball, Jenny H. Pak, Thomas Plante, and Steven L. Porter. 2006. "Invitation to the Table Conversation: A Few Diverse Perspectives on Integration." *Journal of Psychology and Christianity* 25 (4): 338–53.

Young, Emma. 2017. "Psychedelic Drugs and 'Heightened Consciousness.'" *The Psychologist* (journal of the British Psychological Society), July: 17.

Zola-Morgan, Stuart. 1995. "Localization of Brain Function: The Legacy of Franz Joseph Gall (1758–1828)." *Annual Review of Neuroscience* 18: 359–83.

# Name Index

Adam, 25, 134–135, 173, 245, 254, 270–71
Adams, Jay, 74, 91, 245
Agassiz, Jean Louis, 27
Aikman, Duncan, 35, 44, 236, 245
Albright, Carol R., 57, 245
Alexander, Denis, 173–74, 245
Allman, John, 183, 245, 256
Allport, Gordon, 15–16, 63, 152_53, 196, 245
Aquinas, Thomas, 45, 53, 84, 185, 203, 206, 256, 269
Argyle, Michael, 16–17, 245
Aristotle, 35, 45, 53, 84, 203
Arnold, Bill T., 53, 245
Ashbrook, James, 57, 245
Augustine, Saint, 24–25, 39, 49, 52, 84, 185, 202–3, 245, 248, 253

Babbage, Charles, 27, 246
Baddeley, Alan, 167, 246
Barbour, Ian, 33, 78, 240, 246
Barlow, Nora, 27, 246
Barnett, Keri, 87–88, 246
Barr, James, 185, 246

Barrett, Justin, 188–90, 246
Barth, Karl, 186, 220, 246
Bartlett, Frederic, x, 63, 167, 247
Basil, Saint, 45
Basten, Ulrike, 161, 247
Bate, Sarah, 111, 247
Bateson, Melissa, 109, 247
Bauckham, Richard, 167–68, 247, 267
Baumgartner, Günter, x, 258
Beach, Steven R. H., 217, 254
Beauregard, Mario, 140, 269
Beck, James, 59, 69, 83, 94, 247
Bede, Saint, 24
Behensky, Charles, 80, 272
Benovsky, Jiri, 231, 247
Benson, Herbert, 213, 247
Berger, Klaus, 156, 247
Berger, Peter, 157
Berntson, Gary, 154, 250
Berry, R. J., 240–41, 258
Blakeslee, Sandra, 249, 267
Blocher, Henri, 186, 248
Bloom, Paul, 133, 248
Bonhoeffer, Dietrich, 230, 248
Boring, Edwin, 62, 248

Bouma-Prediger, Steven, 74–75, 248
Brahe, Tycho, 26, 47, 248, 256
Brennan, Robert, 52, 248
Broca, Paul, 104, 182, 248
Brooke, John, 32, 43, 48, 50, 78, 248
Brown, Ariel, 146, 263
Brown, Laurence, 17, 248–49
Brown, Warren, 56, 141–44, 164–66, 169, 183–4, 142, 249, 251, 255, 259, 268
Brownson, Kathryn, v
Bruce, Charles, 109, 249
Bruner, Jerome, x, 167, 249
Bunyan, John, 126–29
Burkhardt, Richard, 175, 249
Burns, Jeffrey, 113, 249
Buss, David, 176, 250
Butler, Paul M., 146, 250
Byrd, Randolph, 211–12, 250
Byrne, Michael, 258
Byrne, Richard, 177–79, 250, 271

Caan, Woody, 109, 266
Cacioppo, John T., 154, 250
Call, Josep, 180, 270
Calvin, John, 46–47, 71, 133, 204, 211, 220, 250
Capella Martianus, 25
Carmichael, Amy, 128
Carnegie, Dale, 222, 250
Carson, Donald, 130, 250
Carter, John, 64–66, 69, 250
Castelli, Benedetto, 26
Chemel, Benjamin, 139, 265
Clarke, Henry, 55–56, 251

Clarke, Peter, 57, 251
Clement, Saint, 25
Clements, Ronald, 164, 251
Clinton, Bill, 119, 251
Clinton, Hillary, 119, 251
Coles, Alasdair, 147, 267
Collicutt, Joanna, xii, 96–98, 251
Collins, Gary, 59, 64, 66–69, 92–93, 95, 251
Columbus, Christopher, 22
Combe, George, 55–56, 251
Cooper, John, 51, 135, 251
Copan, Victor, 129–30, 251
Copernicus, Nicolaus, 26, 34, 47, 50, 251
Cosmas (Indicopleustes), 24–25, 46, 49, 251
Cosmides, Leda, 177, 270
Cowan, Charles, 55–56, 252
Cowper, William, 126–28
Crabb, Larry, 59, 69
Cranefield, Paul, 103, 252
Crates (of Mallus), 25–26
Crick, Francis, 101, 252
Crisp, Thomas M., 132, 252

Damasio, Antonio, 114, 117–18, 122–23, 158, 160–62, 165, 252, 256
Darwin, Charles, 12, 27–28, 34, 50, 173, 211, 246, 252, 257, 260
Davies, Gaius, 125–29, 149–50, 252
Davis, Edward B., 43, 252
Davis, Robert, 145–46
Dawkins, Richard, 7

Dax, Marc, 104, 252
de Felipe, Pablo, 22–24, 26, 35,
    45–46, 50, 78, 229, 253
Demarest, Bruce, 69
Descartes, René, 72, 248, 252
Desimone, Robert, 109, 249
De Zuñiga, Diego, 26
de Waal, Frans, 114, 175–76,
    178–79, 253
DeWall, C. Nathan, 4, 219,
    265
di Pellegrino, Giuseppe, 181,
    253
Di Vito, Robert, 156, 253
Dienes, Zoltan, x, 253
Dobson, James, 59, 69
Dostoyevsky, Fyodor, 148, 253
Dowson, Martin, 86, 264
Drake, Stillman, 26, 253
Draper, John, 22, 253
Durso, Raymon, 146, 250, 263

Easton, William, 55–56, 253
Edwards, Jonathan, 36–37,
    204, 253, 263
Eisenberger, Naomi, 163, 254
Elijah, 219
Eliot, T. S., 226
Enns, Peter, 38, 254
Erikson, Erik, 97, 197–99, 208,
    254
Evans, C. Stephen, 70–73,
    84–85, 91, 93–95, 233, 237,
    254
Ezekiel, 219

Farrer, Austin, 208, 254
Feldman, Carol, 167, 249

Fernandez-Duque, Diego, 132,
    254
Ferrier, David, 105, 254
Fincham, Frank, 216–17, 254
Forgas, Joseph, 17, 249
Foscarini, Paolo, 26
Fowler, Orson, 56, 254
Fraser, James, 10
Freud, Sigmund, 6, 11–15, 17,
    58, 61, 70, 97, 197–200, 202,
    206, 208, 254–55
Fritsch, Gustav, 105, 255
Fromm, Erich, 200–1, 206, 255
Furrow, James, 166, 255

Gage, Phineas, 114, 158, 160,
    165, 261–62, 270
Galen, 102
Galileo, 22, 26, 34–35, 195, 253,
    263, 265
Gall, Franz Joseph, 54–55, 102,
    104, 255, 273
Gallese, Vittorio, 182, 253
Gallup Organization, 17, 255,
    259
Galton, Francis, 10, 211, 255
Gardner, Howard, x, 3–4, 255
Geppetto, 223
Gesch, C. Bernard, 115, 255
Gibson, Nicholas, 80–82,
    272
Giese, Tiedemann, 26
Gifford, John, 127
Gilson, Étienne, 203, 256
Goldman, Alvin, 182, 255
Gould, Steven Jay, 48, 236, 256
Grabowski, Thomas, 117–18,
    256

Granada, Miguel, 47, 256
Green, Joel, 27, 39, 51, 132, 135,
     153–159, 169, 187, 245, 256,
     257, 269
Greene, Joshua, 162, 256
Greenway, Tyler, 190, 246
Greer, Brian, x, 259
Groopman, Jeremy, 140, 256
Gross, Charles, 109, 249

Hakeem, Atiya Y. , 184, 245,
     256
Hall, Todd, 83–84, 235, 272
Hamlet, 225
Hare, Brian, 180, 270
Harrison, Peter, 9–10, 21–22,
     32–33, 36, 43, 45, 49, 256,
     271
Hasker, William, 53, 257
Hawkes, Nick, 46, 257
Hearnshaw, Leslie, 10–11, 257
Henslow, John, 27, 246
Hercules, 21
Hitzig, Eduard, 105, 255
Hof, Patrick, 183, 245, 256
Holeman, Virginia, 92, 257
Hopkins, Gerard Manley, 126,
     128, 277
Hubel, David, 100–1, 106, 257
Huxley, Thomas H., 21, 257

Idreos, Andreas, 18, 33
Inge, William, 10
Isaiah, 195, 211, 219, 226
Isidore (of Seville), 24

Jackson, Hughlings, 105, 257
James, 163, 165, 196, 211

James, William, 11–12, 155–158,
     223–24, 257
Jarvinen, Matthew, 189, 246
Jeeves, Malcolm, x, xi, 10,
     22–23, 35, 44, 55–56, 72–73,
     78, 106, 111–12, 122, 153,
     165, 171, 175, 183, 209,
     213, 229–31, 237, 240–41,
     243, 246, 249–50, 252–53,
     256–59, 261–62, 264–66,
     268, 271–72
Jeremiah, 210, 219
Jesus Christ, 38–39, 46, 51, 58,
     62, 68, 112, 129, 133, 141, 153,
     159, 166–68, 186–87, 204,
     215, 220, 223, 247
Job, 163–64, 195, 215
John, 196, 220
Johnson, Dale A., 137, 259
Johnson, Eric L., 3, 5, 29, 31,
     39, 59, 66, 68, 88, 90, 94,
     235, 251, 259–60, 264–67
Johnson, Mark, 119–21 259
Johnson, Micah, 115, 259
Johnson, Samuel, 222, 259
Jones, Paul, 94, 269
Jones, Stanton, 3, 5, 28–29, 40,
     64, 66, 68–69, 73, 89, 90,
     207–8, 212, 251, 259–60, 264
Jung, Carl, 15, 97, 260

Kahn, Chris, 113, 260
Kanwisher, Nancy, 109–10, 117,
     121, 260, 265
Kark, Jeremy, 170, 260
Karmiloff-Smith, Annette, 119,
     260
Kendell, Robert, 123, 260

Keysers, Christine, 110–11, 260
Kierkegaard, Søren, 39, 84,
    220, 260
Kilner, James, 182, 260
Kingsley, Charles, 27, 260
Koteskey, Ronald, 69, 153, 261
Künkel, Fritz, 63–64, 261

Lactantius, 24–25, 51
Lambert, Heath, 69, 261
Lambert, Nathaniel, 217, 254
Langfeld, Herbert, 62, 248
Larner, Andrew, 160, 261
Lashley, Karl, 105, 261
Lassonde, Maryse, x, 165, 261
Leach, John P., 160, 261
Lemaître, Georges, 35, 44, 48,
    236, 245
Lemke, Steve, 233, 261
Lemon, Roger, 182, 260
Lena, Matthew, 160, 262
Leuba, James, 12, 261
Lewis, C. S., 89, 129, 215, 220,
    261
Lewis, Tanya, 138, 261
Lewis Hall, Elizabeth, 82–83,
    95, 235, 261
Lieberman, Matthew, 163, 254
Lindberg, David, 34–35, 261
Loftus, Elizabeth, 167, 261
Lorenz, Konrad, 175, 249
Lucian, 25
Lucretius, 25
Ludwig, Thomas, x, xi, 122,
    209, 213, 259, 262, 272
Luke, 133, 155, 157–58, 218, 256
Luther, Martin, 30, 126–29,
    185, 262

Lyell, Charles, 27

MacKay, Donald, xi, 5, 72–73,
    122, 231, 237–40, 257, 262
Macmillan, Malcolm, 160, 262
Macrobius (Theodosius), 25
Maguire, Eleanor, 118–19, 158,
    263, 272
Maimonides, 170
Malinowski, Bronislaw, 13, 263
Malony, H. Newton, 169, 249,
    261
Marsden, George, 37, 263
Marshuetz, Christy, 121, 266
Marx, Groucho, 221
Marx, Karl, 200
Maslow, Abraham, 92,
    199–200, 202, 206, 208,
    251, 263
Matsui, N., 149, 265
Matthew, 153, 212
Matthews, Dale, 212
McCall Smith, Alexander, 114,
    263
McClendon, Brian, 218, 263
McGrath, Alister, xii, 18,
    33–34, 263
McGrath, Joanna Collicutt,
    xii, 96–98, 251
McKnight, Scot, 173, 270
McMullin, Ernan, 27, 50, 263
McNamara, Patrick, 146, 250,
    263, 265
Meehl, Paul, 30–31, 87, 91, 263
Middleton, J. Richard, 186, 264
Milner, A. David, x, 264, 266
Miner, Maureen, 86, 264
Miskin, Prince, 148

Mivart, George, 27
Moes, Paul, 29, 153, 259, 264
Mohline, Richard, 65–66, 250
Moltmann, Jürgen, 187
Moses, 13, 46–47, 210, 250, 255
Munn, Norman, 62, 264
Murphy, Nancey, 169, 249
Myers, David, xii, 4–8, 39–40, 73, 88–89, 153, 170, 209–210, 212, 219, 227, 237–38, 241, 264–65

Naito, H., 149, 265
Narramore, Bruce, 65, 69, 82, 91, 250, 261
Narramore, Clyde, 58, 69
Nettle, Daniel, 109, 247
Newberg, Andrew, 159
Nichols, David, 139, 265
Noah, 25
Norlie, Olaf, 64, 265
Norman, Wayne, 55, 259, 265
Numbers, Ronald, 24, 32, 24–35, 49–50, 261, 265

Obama, Barack, 101
O'Craven, Kathleen, 117, 121, 265
Owen, Richard, 27

Palmer, John, 167, 261
Pannenberg, Wolfhart, 169, 187, 265
Pascal, Blaise, 176, 220, 225, 266
Paul, 40, 60, 126, 131, 194, 196, 210, 215, 219, 222–23, 269
Paul, Lynn, 184, 279

Pavlov, Ivan, 105, 154, 266
Payer, Doris, 121–22, 266
Penchansky, David, 163, 266
Perdue, Charles W., xi, 262
Perrett, David, 109–11, 260, 266
Petrovic, Predrag, 116, 266
Phillips, J. B., 128, 279
Philoponus, John, 24, 45–46
Phinney, Raymond, 135, 268
Photius, Saint, 24
Piccolomini, Ascanio, 26
Pinocchio, 223
Plato, 25, 45, 50–53, 131, 231
Plutarch, 25
Pope Paul III, 26
Porter, Steven, 132, 253, 273
Powell, Baden, 27, 266
Powlison, David, 88, 266
Pratt, James, 11, 266
Premack, David, 177, 266
Prentice-Dunn, Steven, 218, 263
Ptolemy, 25, 48

Rabin, John, 149, 268
Rahner, Karl, 205, 207, 266
Rajalakshmi, R., xi, 175, 266
Ramachandran, Vilayanur S., 57, 149, 181, 267
Ramakrishnan, Venki, 49, 112, 234, 267
Randles, William, 25, 267
Redfern, Clare, 146–47, 267
Redman, Judith, 167, 267
Rees, Dai, 123, 263, 267
Rizzolatti, Giacomo, 181–82, 253

Roberts, Gilbert, 109, 247
Roberts, Robert, 5, 39, 74, 84,
    88, 267
Rogers, Carl, 199–200, 202,
    206, 268
Rolls, Edmund, 109, 266
Rossetti, Christina, 126, 128
Rothmann, Christoph, 47,
    256
Rugg, Michael, x, 264
Rumbaugh, Duane, 175, 268
Russell, Colin, 31–32, 268

Sadato, Norihiro, 118, 268
Salzman, Mark, 148, 268
Samples, Kenneth, 233, 261
Samuel, 195, 245
Sargant, William, 16, 153, 268
Saver, Jeffery, 149, 268
Schloss, Jeffery, 164, 268
Scott, Rodney, 135, 268
Scott, William, 55–56, 268
Sedgwick, Adam, 27
Shaftesbury, Lord, 126
Shirakatsi, Anania (Anania of
    Shirak), 24
Siegel, Zachary, 139, 268
Sinkman, Arthur, 162, 268
Skinner, B. F., 7, 12, 14–15, 17,
    61, 70, 269
Smedes, Lewis, 136–37, 269
Sperry, Roger, 107–8, 117, 269
Spiers, Hugo, 119, 272
Spinks, G. Stephens, 15, 269
Spurzheim, Johann, 54–55,
    104, 269
Starbuck, Edwin, 10, 269
Stawski, Christopher, 140, 269

Stephen, 133
Stone, Lawson, 134–35, 269
Stott, John, 126, 194, 205, 269
Strawn, Brad, 94, 141–44, 183,
    242, 249, 269
Sullivan, J. E., 202, 269
Swerdlow, Russell, 113, 249

Tellinghuisen, Donald, 29, 153,
    264
Temple, William, 152
Ten Elshof, Gregg, 132, 252
Tertullian, 51–52, 248
Thiselton, Anthony, 131, 134,
    187, 243, 269
Thomas, Gary, 212, 270
Thomson, Keith Stewart, 215,
    279
Thorpe, William, 174, 270
Thouless, Robert, 11–12, 270
Tinbergen, Nikolaas, 175, 249
Tjeltveit, Alan, 87, 91, 270
Tomasello, Michael, 179–80,
    189, 270
Tooby, John, 177, 270
Tournier, Paul, 69
Tree, Jeremy, 111, 247

Vande Kemp, Hendrika, 63,
    69, 79, 270
Vanderlaan, Kelly, xi, 272
Van Horn, John, 114, 270
Van Leeuwen, Mary Stewart,
    82, 91, 235, 261
Venema, Dennis, 173, 270
Vesalius, 103
Vitz, Paul, 84, 207, 270
Volf, Miroslav, 131, 271

von Middelburg, Paul, 26
von Rad, Gerhard, 186, 271
von Schoenberg, Nikolaus, 26

Wagener, Linda, 166, 255
Walton, John, 28, 37–38,
    232–33, 261, 271
Watson, P. J., 5, 39, 74, 88, 267
Watts, Fraser, 9, 18–19, 77, 81,
    85–86, 241, 271
Weaver, Glenn, 145, 271
Weld, Harry, 62, 248
Wells, David, 154, 271
Wells, Gary, 168, 272
Wernicke, Carl, 104, 271
Wesley, Charles, 126
Wesley, John, 126
Westerholm, Barbro, 123, 267
Westermann, Claus, 187, 271
White, Andrew, 22, 271
Whiten, Andrew, 177, 183, 188,
    250, 271
Wiesel, Torsten, 100, 106, 257
Williams, Kipling, 163, 254
Willis, Thomas, 103, 252, 272
Winefield, Anthony, xi, 259

Witvliet, Charlotte vanOyen,
    xi, 272
Wixted, John, 168, 272
Wolska-Conus, Wanda, 24,
    272
Wolterstorff, Nicholas, 69, 272
Woodruff, Gary, 177, 266
Woodward, Kenneth, 212, 272
Woollett, Katherine, 119, 272
Worthington, Everett, 9,
    29–30, 40–41, 57–60, 83–85,
    89–90, 93, 235, 241, 272
Wright, Christopher, 186, 272
Wright, David, 258
Wright, N. T., 37, 130–32, 134,
    272
Wright, Ronald, 94, 269

Yangarber-Hicks, Natalia,
    79–81, 272
Young, Emma, 130, 273
Yovel, Galit, 110, 260

Zola-Morgan, Stuart, 102–3,
    273

# Subject Index

accommodation theory, 45–48
adolescence, 15, 166, 198
agency, 92, 158, 186
alien hand syndrome, 159
Alzheimer's disease, 102, 144–45, 149, 242
anterior cingulate cortex (gyrus), 118, 163, 183
anthropology, 9–10, 13, 50–53, 85, 96–97, 135, 169, 202, 231
anthropology, dualistic, 27, 50–53, 72, 101, 122, 131–35, 141–42, 160, 169, 231, 237
anthropomorphism, 97
anticipatory evaluative responses, 161–62
antidepressants, 128, 150
antipodeans, 23, 25, 49
antisocial behavior, 114–15, 199
anxiety, 57, 126–28, 148, 163, 196, 203, 238
apes, great apes, 182–84, 186
atheism, 22, 212
attitudes, 14, 29, 31, 45, 62–63, 95, 116, 152, 164, 218–20, 226–27, 237
attribution, 97, 178–79, 221

authentic dialogue (convergence), xii, 209, 211, 220, 241
autistic spectrum disorder, 177
automaticity of behavior, 182, 210, 225
autonomic response system, 161
axons, 120, 138, 183
ayahuasca plant, 137, 139

behaviorism, 14–15, 106–7
biblical counseling, 74, 88, 91
biblical interpretation, 7, 9, 35–38, 40–41, 43, 53, 60, 65, 133–35, 195, 230–34

body-mind issue, 72, 104, 116, 141–43, 146, 149
body-soul dualism, 27, 50–53, 72, 101, 131–35, 141–42, 160, 169, 231, 237, 242
bottom-up processing, 80, 103, 107, 117
brain damage (dysfunction), 102, 104–5, 107–8, 111, 113–15, 131, 151, 158, 160–63, 165–66, 169, 242
brain development, 110, 119–21, 158, 166

brain hemispheres, xi, 6, 104, 107–8, 146, 165

brain imaging (scanning, mapping), 104, 106–9, 113, 116–19, 121, 136, 161–63, 183

brain-mind link, 72, 104–6, 116, 141–43, 146, 149

brain specialization (localization of function), xi, 6, 102–13

brainstem, 116

brain tumor, 105, 113, 211

Capgras syndrome, 159, 162–63

cerebral cortex, 54, 105, 120–22, 137–38, 160–63, 165, 178, 182–84

cerebral hemispheres, xi, 6, 104, 107–8, 146, 165

character, 54, 64, 114, 134, 151, 164, 218

characteristic, 19, 111, 134–35, 158, 175, 177–78, 186, 191, 195, 201

child (child development), 6, 14–15, 69, 110–13, 119, 133, 158, 198, 210, 215–16, 225

chimpanzee, 180, 188

Christian counseling, 58–59, 63, 74, 78–79, 88–89, 91–92, 94, 97, 218

Christian psychology, 5–6, 39, 59, 64, 74, 79, 81, 83–84, 88–90, 95, 235

church attendance, 63, 147

church history, 35, 37, 45, 125, 202

church (congregation), 41, 60, 62–63, 98, 140–44, 166, 242

church (institution), 26, 30, 35, 37, 45, 50–52, 102, 125, 133, 202, 205, 229, 231

cingulate cortex (gyrus), 116, 163, 183

client-centered therapy, 200

clinical (counseling) psychology, 50, 57, 59, 74, 78–79, 81–83, 86, 88–89, 92, 94, 97, 99, 173, 193, 196, 233

cognitive capacities (functions), 6, 107–8, 116, 121–22, 145, 165–66, 169, 178, 182–85, 191, 224–25

cognitive neuroscience, x–xi, 3–4, 80, 106, 117

cognitive psychology (cognitive science), x–xi, 3, 17, 19, 73, 97, 99, 106, 111, 154–55, 157–59, 168, 188–90, 233, 241, 243

cognitive revolution, x, 3, 106, 154

cognitive science of religion, 188–90

collective unconscious, 15

community of faith, 58, 98, 140, 152, 159, 171, 242

comparative psychology, xi, 106, 175, 180

compartmentalism, 236

compassion, 11, 115, 162, 197, 242

compatabilists, 70–71

complementarity (complementary perspectives), x–xi, 8–9, 18–19, 30, 70, 72–74, 80–81, 85–86, 95, 119, 199, 205–6, 219, 222, 225–29, 237–241

complex systems, 113, 137, 175, 180

concordism, ix–xii, 7, 10, 12–13, 16, 19, 33, 35–36, 39, 43–45, 47–50, 52–56, 58–61, 84, 90, 112, 150, 167, 174, 178, 180, 194, 196, 209,

211, 216, 227, 230–233, 236–237, 241

conflict (science-religion), 21–23, 25, 27–36, 41, 43–44, 150, 174, 211, 229

consciousness, 11, 15, 70, 80–81, 100, 104, 107, 117, 122, 130, 135, 139, 148, 160–61

constructive relationship (science-religion), ix–x, 32–33, 68, 93, 150, 229, 233, 236

control beliefs, 69

conversion, 11, 16–17, 29, 126, 151–59, 171, 204

corpus callosum, 165

culture, 38, 40, 114, 129, 133, 179

cyclothymia, 126

Darwinism, 12, 27–28, 50, 173

Decade of the Brain, 4, 101

Decade of the Mind, 4, 101

decision-making, 16, 100, 106, 108, 113, 115, 151, 160–62, 165, 169, 183, 185

deep social mind, 183, 188

dementia, 102, 144–145, 149, 242

depression, 11, 57, 126–28, 130, 210, 223

depth psychology, 78, 97

determinism, 71, 107

developmental psychology, 69, 98, 111, 120, 133, 140, 166, 208

dialogue, xii, 18, 29, 30, 33, 66, 81, 85, 209

domains of knowledge, 8, 19, 22, 32, 43, 64, 68, 72, 74, 130, 137, 208, 234, 236, 239

dopamine, 138, 146

drugs, 126–27, 130, 136–39

dual-aspect monism, 231, 237

dualism, 27, 50–53, 72, 101, 131–35, 141–42, 160, 169, 231, 237, 242

dualistic anthropology, 27, 50–53, 72, 101, 122, 131–35, 141–42, 160, 169, 231

duality, 122, 231

ecstatic experiences, 129, 137, 147, 149

ego, 196–99, 208

embedded (embeddedness), 125, 140–143, 156, 170–71, 195

embodied (embodiment), 92, 103, 114, 125, 132, 136, 138–39, 141–43, 149, 157–58, 162, 169–70, 242

emergent properties, 108, 117, 143, 161, 187, 191

emotions (emotional), 11, 16, 57, 98–100, 109, 136, 142, 145, 147, 149, 152, 154, 161, 163, 183–84, 217

empathy, 154, 162, 183

empiricism, 16, 25–26, 31, 40, 83, 89, 210, 215

enrichments (science-religion), x, xii, 7–9, 18–19, 30, 34, 80, 86, 96, 151, 154, 190, 194, 206, 229, 241

epilepsy (seizures), 57, 105, 117, 136, 147–148

establishment (mainstream) psychology, 6, 40, 59, 74, 77, 89, 96–97

Eucharist, 130, 141

evangelical Christianity, 64, 174, 223

evangelism, 16
event-related potential, 119
evil, 9, 57, 199–201, 203
evolution, 22–23, 27–28, 98, 135,
    174–75, 177, 181–82, 190, 193, 232
evolutionary psychology, x–xi,
    3–4, 19, 41, 73, 81, 87, 98, 173–177,
    179–182, 184–188, 190–91, 193,
    196, 233, 241–243
executive processes, 113, 198
exocentricity, 187
explaining away, 12, 17, 150, 152–53,
    207
eyewitness memory, 6, 152, 166–68,
    171

face-blindness, 111
face perception (processing),
    108–12, 117, 120–21
faith, 11–17, 28, 55–56, 63, 71, 73, 75,
    125–26, 141, 144–45, 152, 167, 170,
    212, 214–15, 219–220, 223, 226
faith healing, 75, 170, 212–14
father figure, 12–15
fear, 196, 203
feelings (emotions), 129, 139, 141,
    149, 155–56, 161–63, 171, 179,
    196–97, 218, 222
feelings of familiarity (love), 109,
    159, 162–63
flat earth, 22–25, 46, 49
free will, 14, 51, 70–71, 102. 113, 115,
    201, 205
frontal lobes (cortex), 104, 113–14,
    121–22, 146, 158, 160–63, 165, 184,
    242
  orbitofrontal cortex, 160–61, 165
  prefrontal cortex, 122, 146, 184

functional magnetic resonance
    imaging (fMRI), 107, 109, 121,
    136, 161–63
fusiform face area, 109–10, 117

gaze, direction of, 109
generativity (Erikson), 199
genes (genetic influences), 19, 110,
    115, 119–20, 174–75, 178–79, 190,
    193, 196
Genesis, 28, 46–47, 133–35, 185, 187,
    190
geological knowledge, 195
gestures, 141, 224
glossolalia, 10, 159
glutamate, 137
goal-directed behavior, 80, 146,
    182
God, 7, 12, 14–17, 20, 30, 35, 37–38,
    44–48, 52, 54–55, 57, 64–66,
    68–69, 71, 75, 86, 92–93, 95, 118,
    123, 126, 133–34, 137, 140–41, 143,
    145, 147–49, 154, 158–59, 164,
    170–71, 175–76, 178, 180, 182,
    185–190, 194, 202–209, 211–12,
    214–16, 222–26, 230, 232–33,
    235–36, 239, 241, 243
God module, 149
god of the gaps, 35, 69, 71, 214, 239
God spot, 57, 232
grace, 75, 155, 202–4, 206, 222–23,
    226
Greek thought, 45, 49, 50–53, 102,
    131–32, 231
groups (group identity), 12, 16, 98,
    154, 163, 167, 200–1, 222
guilt, 9, 11, 13, 115, 145, 179, 196, 200,
    205, 207

hallucinations, 105, 138–39

hallucinogenic (psychedelic) drugs, 130, 136–39

health (healing), 11, 75, 81, 170–71, 197, 212, 214–15, 217, 221, 223

heart, 51, 100, 102, 129, 141, 148, 161, 196, 203–4, 212, 217–18, 220

Hebrew thought, 38, 47, 53, 131, 134, 169, 220, 231

hemispheres, xi, 6, 104, 107–8, 146, 165

hemispheric specialization (differences), xi, 6, 104, 107–8, 146

heritability (inherited traits), 6, 178, 190, 193

hermeneutics, 7, 9, 35–38, 40–41, 43, 53, 60, 65, 133–35, 195, 230–34

hierarchy of needs, 199, 208

higher-order theory of mind, 189–90

hippocampus (and memory), 117–19

Homo sapiens, 183, 225

humanistic psychology, 199–200, 202, 206, 208

human nature, 7–9, 39, 41, 43, 50, 55–56, 66, 69, 74, 81, 84–85, 93, 98, 100, 118, 124, 131, 134, 142, 155, 171, 173–75, 178, 188–89, 194–97, 199, 201–4, 206–8, 218, 226, 230, 234, 242, 244

human needs, 12, 14, 140, 193, 195–97, 199, 201–8, 215–16

humanness (human dignity), 14, 175, 183, 185, 187, 209, 242

human uniqueness (distinctiveness), 11, 103, 134–35, 176–78, 180–89, 201, 242–43

id, 196–98, 208

identity, 12, 156, 163, 187, 198, 201

*Idiot* (Dostoyevsky), 148

illusion, 13, 225, 225, 236

image of God (*imago Dei*), 134, 175, 180, 182, 185–90, 224–26, 243

individual differences, 8, 11, 111, 144–47, 155–56, 160, 170–71, 177–79

individualism, 155–56

information processing, 6, 97, 138, 165, 175, 179, 183–85, 225

inherited traits, 6, 178, 190, 193

Inquisition, 26

insights (science-religion), x, xii, 7–9, 12, 16–19, 27, 30, 33, 37, 39, 59, 80, 96, 98–99, 114–115, 123, 125, 132–33, 135–36, 139, 141, 144, 146–47, 150–51, 154–55, 157, 164, 166–69, 171, 173, 180, 185, 187–88, 190, 193–95, 206–9, 229, 232, 241–43

insula (insular cortex), 183–84

integration, xi–xii, 8, 19, 33, 35, 40, 44–45, 58–61, 63–71, 73–75, 77–95, 97, 194, 206–7, 232–37

intelligence, 6, 19, 118, 160, 164–65, 178, 210

interaction, 32–33, 106, 113, 119, 122, 143, 156, 162–64, 197, 220

interdependence, 100–1, 122

interpersonal relations (relatedness), 92, 162, 187–89, 243

interpretation (biblical), 7, 9, 35–38, 40–41, 43, 53, 60, 65, 133–35, 195, 230–34

intimacy (Erikson), 198, 215, 224

intrinsic, 65, 122, 223

intuition, 6, 156, 161

kibbutz, 170–71

language, 6, 8–9, 15, 29, 37, 43,
46, 63–64, 72, 74, 95, 104, 106,
134, 164, 179, 181–82, 184, 188,
194–96, 206, 208, 219, 224, 226,
233–36, 242
learning, x, 14, 74, 82, 151, 174–75,
178, 180, 185, 235, 242
levels of explanation, x–xi, 8–9,
18–19, 30, 70, 72–74, 80–81,
85–86, 95, 119, 199, 205–6, 219,
222, 225–29, 237–241
limbic cortex (system), 149, 161–63,
183
localization of function, xi, 6,
102–13
locus coeruleus, 138
Lord's Prayer, 214

magnetic resonance imaging
(MRI), 107, 113, 118
mainstream (establishment)
psychology, 6, 40, 59, 74, 77, 89,
96–97
mechanistic processes, 26, 29, 69,
71, 103, 108–10, 116, 137, 153–54,
178
meditation, 57, 171, 214
memory, 6, 13, 97, 99, 106, 138–39,
159, 166–71, 178, 180, 224, 226
memory reliability, 166–71
midbrain, 138
mind, 4, 16, 19–20, 39, 47, 53–54,
56, 67, 72, 92, 99–101, 104–7,
111, 116–17, 122–23, 129, 133, 135,

141–43, 145–46, 148–49, 153, 158,
169, 174, 176–83, 188–90, 196,
210–11, 218–19, 230–31, 237
mind-body problem, 72, 104, 116,
141–43, 146, 149
mind-brain link, 72, 104–6, 116,
141–43, 146, 149
mind-reading (theory of mind),
177–78, 180–82, 188–190, 211
mirror neurons, 181–82
monism, 27, 53, 101, 131–32, 135, 160,
169, 231
monist anthropology, 53, 135, 169,
231
monkeys, 105–6, 109–10, 118, 120,
178, 182, 184, 188
monks, 23, 46, 127
moral action (behavior), 13, 17, 102,
114–15, 159, 161, 178, 193, 200,
218–19, 242
moral (morality), 13, 17, 51, 55, 100,
102, 114–15, 151, 159–65, 169, 171,
178, 183–84, 193, 198, 200, 207,
210, 218–19, 222, 236, 242
moral responsibility, 51, 114–15, 169,
242
moral decision-making, 100, 102,
151, 161–62, 165, 169, 171, 183–84,
210, 242
morphine, 116
motivation, 22, 97, 146, 193, 196,
199, 201–3, 217, 223
motor cortex (centers), 105, 109,
121, 182
multiple levels of explanation, x–xi,
8–9, 18–19, 30, 70, 72–74, 80–81,
85–86, 95, 119, 199, 205–6, 219,
222, 225–29, 237–241

mutual insights (enrichments), x, xii, 7–9, 12, 16–19, 27, 30, 33, 37, 39, 59, 80, 96, 98–99, 114–115, 123, 125, 132–33, 135–36, 139, 141, 144, 146–47, 150–51, 154–55, 157, 164, 166–69, 171, 173, 180, 185, 187–88, 190, 193–95, 206–9, 229, 232, 241–43
mystery, 87, 123, 128, 207
mysticism, 10–11, 136–37, 140

natural-born dualists, 133
naturalistic explanation, 12, 200
natural philosophy, ix, 43, 45, 48, 50, 230–31
natural selection, 22–23, 27–28, 98, 135, 174–75, 177, 181–82, 190, 193, 232
nature vs. nurture, 6, 118–22
needs (human), 12, 14, 140, 193, 195–97, 199, 201–8, 215–16
*nephesh* (soul), 134–35
Nestorianism, 24
neural activity, 6, 69, 80, 103–6, 109, 111, 118, 120–22, 136–37, 140, 146, 149, 154, 159, 166, 181–82, 224
neural networks, 6, 80, 104–5, 115–16, 122, 136–37, 140, 146
neural plasticity, 107, 110, 118–22
neuropharmacology, 8
neuropsychology, x, xi, 3, 8, 19, 54, 77, 87, 98–100, 102, 106–8, 112, 125, 131, 139, 141, 144, 150, 154–55, 164–66, 171, 173, 181, 193, 196, 233, 241, 242
neuroscience, x, xi, 3–4, 19, 27, 41, 53, 56–57, 70, 80, 92–93, 00–102,

105–6, 111, 113–14, 116–17, 123, 129, 132–33, 135–36, 140–41, 149, 151, 154–55, 158–59, 161, 163, 165, 171, 181–83, 185, 231
neurotheology, 54, 56, 136, 140, 231
neurotic, 126, 137
neurotransmitters, 115, 137–38
New Testament, 38, 51, 135, 153–54, 156, 159, 166–67, 206
nonhuman primates, 106, 177–82, 189, 243
nonoverlapping magisteria, 48, 66, 236–37
non-reductive physicalism, 142
nothing buttery (reductionism), 9, 15, 18–19, 150, 178, 236
novelty, perception of, 138–39

obsessive-compulsive disorder, 127–28
Oedipus complex, 12
Old Testament, 38, 53, 134, 156, 232
open-minded, 41, 180, 209, 215, 243
orbitofrontal cortex, 160–61, 165
origins of religion, 10, 12–13, 15, 166
out-of-body experience, 139, 210–11

parahippocampal area, 117
parietal lobes (cortex), 57, 162
Parkinson's disease, 146–47, 242
pedophilia, 113–14
person (personhood), 9, 11, 51, 53, 70–73, 101, 131–35, 142–43, 156, 160–65, 187, 200, 205, 207, 231, 237, 241
personality, 8, 11, 54, 98–99, 106, 112, 126, 128, 142, 158, 195–99, 201, 205–6, 233

personality development, 12–15, 197–201, 210

personality traits, 6, 19, 54, 134

personal relatedness, 92, 162, 187–89, 243

perspectivalism, x–xi, 8–9, 18–19, 30, 70, 72–74, 80–81, 85–86, 95, 119, 199, 205–6, 219, 222, 225–29, 237–241

persuasion, 218–19

philosophy, ix, 32, 43, 48, 64, 70, 132, 199–200, 203, 207, 231

phrenology, 7, 50, 54–56, 103, 231–32

physicalism, 143, 231

placebo analgesia, 116

plasticity (neural), 107, 110, 118–22

positive psychology, 223

positron emission tomography (PET), 107, 116

prayer, 10–11, 130, 143, 171, 211–17

prayer effects (benefits), 211–17

prayer experiment, 211–17

prefrontal cortex, 122, 146, 184

presence of God, 140, 143, 147, 203

presuppositions, 15, 39, 51, 90, 94, 194, 207

pride, 9, 221–24, 226–27

primates, 109–10, 178–83, 188–89, 243

Prince Miskin (Myshkin), 148

problem-solving, 164–65, 180, 184

prosocial behavior, 15

prosopagnosia, 111

Prozac, 137

psychedelic experiences (drugs), 130, 136–39

psychiatric illness, 126–27, 149, 202

psychiatry, 16, 87, 105, 113, 122–23, 125, 128, 149, 153

psychoanalysis, 13–15, 197, 208

psychobiological unity, 51, 100, 131, 231, 241–42

psychology of religion, 10–11, 15, 17, 151, 153–54

psychology-religion models, 6–20, 30, 41, 45–47, 63, 65–66, 70–74, 207–8, 237–39

conflict–retreat model, xi, 14, 17, 19, 21–36, 41, 43–44, 50, 60, 150, 173–74, 211, 229–30, 237

filter model, 30, 58, 69

integration model, xi–xii, 8, 19, 33, 35, 40, 44–45, 58–61, 63–71, 73–75, 77–95, 97, 194, 206–8, 232–37

perspectival (insights-enrichments) model, x–xi, 7–10, 12, 16–19, 27, 30, 33, 37, 39, 59, 70, 72–74, 80–81, 85–86, 95–96, 98–99, 114–115, 119, 123, 125, 132–33, 135–36, 139, 141, 144, 146–47, 150–51, 154–55, 157, 164, 166–69, 171, 173, 180, 185, 187–88, 190, 193–95, 199, 205–9, 219, 222, 225–29, 232, 237–243

relational model, 30, 33

psychopathology, 87, 126–27, 149, 202

psychophysics, 93

psychosomatic unity, 51, 100, 131, 231, 241–42

psychotherapy, 57–59, 78–79, 86, 94, 200, 207

psychotic behavior, 127

psychotropic drugs, 127, 136

raphé nucleus, 138
rationality, 14, 22, 55, 103, 185, 225, 227
reality, 9, 11–13, 30, 33, 57, 67–68, 70–73, 85, 117, 135, 139–40, 149, 157, 163–64, 170, 201, 226
reason (reasoning), 22, 53, 55, 66, 82, 102, 151, 169, 171, 190, 210, 225, 242
reciprocal connections (interactions), 106, 113, 119, 142–43, 149, 163–64, 197, 220
redemption, 194, 202–4, 244
reductionism, 9, 15, 18–19, 150, 178, 236
relational (relationality), 30, 92, 187–89, 243
religion, 10–17, 74–75, 153–59, 170, 194–96, 199, 236
  cognitive science of, 188–90
  primitive vs. developed, 11–14
  psychology of, 10–11, 15, 17, 151, 153–54
religiosity (religiousness), 11–12, 17, 136, 146–49, 155
religious beliefs, 7, 10, 16–17, 28, 44, 55–56, 58–59, 68–69, 73, 93–94, 101–2, 112, 139, 146, 150, 152–53, 166, 169–71, 173, 207, 229–30, 236
religious behaviors, 10, 17, 146–47
religious conversion, 11, 16–17, 29, 126, 151–59, 171, 204
religious development, 11–16, 140
religious experiences (states), 11, 16–17, 56–57, 136–40, 147–49, 152–59, 169, 241–42

ecstatic states (drugs), 16, 136–40, 147–49
  meditation, 57, 171, 214
  seizures, 57, 105, 136, 147–49
responsibility, 51, 169, 221
resurrection of the dead, 133–34, 237
royal priesthood, 134

sacred, 47, 97, 152, 204
scientific enterprise, 28, 32, 39, 61, 150
scripture, xii, 7–8, 37–41, 44–46, 54, 59–60, 64–66, 68–69, 93, 166–68, 194–96, 210, 222, 232–34
self-acceptance, 223–24, 226
self-actualization, 199–200, 202
self-awareness (self-consciousness), 135
self-efficacy, 223
self-esteem, 221–23, 227
selfish genes, 19, 179
self-persuasion, 219
self-serving bias, 221–22, 224, 226–27
self-transcendence, 187
serotonin, 137–38
sexual orientation, 210
shame, 183, 185, 242
social behavior (functioning), 106, 154, 162–65, 175, 178–79, 181, 184, 188
social cognition, 110–12, 142, 163, 178–79, 181, 183
social context (environment), 16, 38, 140–43, 154, 170–71, 184, 191, 193, 197
social identity, 156, 201

social inclusion (exclusion), 163
social intelligence, 154, 160, 165
social neuroscience, 154, 181–85
social psychology, xi–xii, 8, 19, 98,
    141, 193, 209, 218–20, 222, 225,
    233
sociology, 70, 96, 157–59
somatic marker theory, 161–62
soul, 39, 50–53, 55, 67, 72, 99–103,
    131–35, 141–44, 146, 149–50, 160,
    169, 171, 186, 231, 242
sphericity of the earth, 22–25, 46,
    49
split-brain patients, 108
spirit, 11, 39, 45, 52, 67, 100, 133,
    142–43, 149, 159, 196, 204, 208,
    216
spiritual healing, 75, 170–71, 211–17
spiritualism, 210–11
spirituality, 11, 15, 17, 56–58, 71–72,
    86, 102, 125, 129–31, 149–50, 159,
    169–71, 204, 214–15, 219, 241–42
  embedded, 125, 140–44, 170–71
  embodied, 125–29, 131–40, 169,
    241–42
  impaired, 125, 144–49
stroke, 155
subjective experience, 115–16,
    137–39, 156
symbols, 12, 15, 181, 196
synapses, 120

temporal lobe, 57, 105, 109, 147–49,
    162–63, 242
temporal lobe epilepsy, 57, 147–49
thalamus, 138

theism, 209
theodicy, 164
theological presuppositions, 15, 39,
    51, 90, 94, 194, 207
theology, xii, 7, 9–10, 18–19, 36–38,
    40–41, 44–49, 51, 65–67, 74, 81,
    87, 131, 223, 236, 239
theory of mind (mind-reading),
    177–78, 180–82, 188–190
top-down processes, 80, 104, 107,
    115–17, 136–37
totem, 12–13
transcendence, 12, 15, 187, 199,
    205–6, 223
two books metaphor, 45, 48

unconscious conflicts, 12, 198
unconscious processes, 11–12,
    14–15, 161, 198, 201, 210
uniqueness (human), 11, 103,
    134–35, 176–78, 180–89, 201,
    242–43

ventral tegmental area (midbrain),
    138
visions, 147–48
vocation, 74, 134, 189, 243
von Economo neurons, 181–84

warfare metaphor, 21–23, 25, 27–36,
    41, 43–44, 150, 174, 211, 229
Weber's law, 93
wisdom, 151, 163–66, 169, 171, 184,
    222
wisdom literature, 163
worship, 141, 143–44, 147, 211, 242